Modelling, Monitoring and Management of

Forest Fires II

WIT*PRESS*

WIT Press publishes leading books in Science and Technology.
Visit our website for new and current list of titles.
www.witpress.com

WIT*eLibrary*

Home of the Transactions of the Wessex Institute.
Papers presented at Forest Fires 2010 are archived in the WIT eLibrary in volume 137 of WIT Transactions on Ecology and the Environment (ISSN: 1743-3541).
The WIT eLibrary provides the international scientific community with immediate and permanent access to individual papers presented at WIT conferences.
Visit the WIT eLibrary at www.witpress.com.

SECOND INTERNATIONAL CONFERENCE ON
MODELLING, MONITORING AND MANAGEMENT OF

Forest Fires 2010

CONFERENCE CHAIRMEN

G. Perona
Politecnico di Torino, Italy

C.A. Brebbia
Wessex Institute of Technology, UK

INTERNATIONAL SCIENTIFIC ADVISORY COMMITTEE

Organised by
Wessex Institute of Technology, UK
Politecnico di Torino, Italy

Sponsored by
WIT Transactions on Ecology and the Environment

WIT Transactions

Transactions Editor

Carlos Brebbia
Wessex Institute of Technology
Ashurst Lodge, Ashurst
Southampton SO40 7AA, UK
Email: carlos@wessex.ac.uk

Editorial Board

Modelling, Monitoring and Management of

Forest Fires II

Editors

G. Perona
Politecnico di Torino, Italy

&

C.A. Brebbia
Wessex Institute of Technology, UK

WIT*PRESS* **Southampton, Boston**

G. Perona
Politecnico di Torino, Italy

C.A. Brebbia
Wessex Institute of Technology, UK

Published by

WITPress
Ashurst Lodge, Ashurst, Southampton, SO40 7AA, UK
Tel: 44 (0) 238 029 3223; Fax: 44 (0) 238 029 2853
E-Mail: witpress@witpress.com
http://www.witpress.com

For USA, Canada and Mexico

Computational Mechanics Inc
25 Bridge Street, Billerica, MA 01821, USA
Tel: 978 667 5841; Fax: 978 667 7582
E-Mail: infousa@witpress.com
http://www.witpress.com

British Library Cataloguing-in-Publication Data

A Catalogue record for this book is available
from the British Library

ISBN:978-1-84564-452-9
ISSN: 1746-448X (print)
ISSN: 1743-3541 (on-line)

The texts of the papers in this volume were set individually by the authors or under their supervision. Only minor corrections to the text may have been carried out by the publisher.

No responsibility is assumed by the Publisher, the Editors and Authors for any injury and/or damage to persons or property as a matter of products liability, negligence or otherwise, or from any use or operation of any methods, products, instructions or ideas contained in the material herein. The Publisher does not necessarily endorse the ideas held, or views expressed by the Editors or Authors of the material contained in its publications.

Printed in Great Britain by MPG Books Group, Bodmin and King's Lynn.

Preface

This book contains peer-reviewed papers presented at the Second International Conference on Modelling, Monitoring and Management of Forest Fires held in Kos, Greece, in 2010. The papers covered important topics in the field of prevention and fighting of forest fires. The Conference was organised the Wessex Institute of Technology of the UK in collaboration with the Politecnico di Torino, Italy.

As in the past, future forest fire scenarios are impacted by climatic trends and changes in climatic extremes, as well as by anthropic pressure. It is to be expected that future trends, especially in the Mediterranean regions, will certainly lead to an increasing impact of human pressure on the natural environment, due to increases in tourism and to the enlargement of urban residential areas invading the countryside.

Forecasting the effects of both factors (climatic and anthropic) and separating their effects on forest fires frequencies may be particularly difficult, but is essential to improve our knowledge of forest fire occurrence probability and to better organize prevention and fighting activities. At the same time, estimation of the possible increase of fire risk over coming years is important, taking into account also the diverse fire prone environments present in many areas of the world.

Although in the majority of cases fire onsets are due to negligence or arson, it is well known that meteorological parameters are extremely important in determining fire risk. Presently, the Joint Research Centre in Ispria, Italy is publishing a daily bulletin for fire danger forecast in Europe using as input data the output of models of the European Centre for Medium Range Weather Forecasts. However, due to the complex orography of most regions, a noteworthy improvement could be reached by using high resolution weather forecasts in conjunction with a detailed description of the configuration. Furthermore, high resolution meteorological fields (mainly wind field) description, in connection with a detailed orographic representation, is essential in predicting fire propagation behaviour, which in turn provides extremely valuable knowledge for any direct activity on the fire itself.

It can be noted that all over the world "uncontrolled vegetation fires contribute to global warming, air pollution, desertification and loss of biodiversity. Between 2000 and 2009, over 200,000 fires have been reported in Sudan and 400,000 in

Ethiopia, for instance." At the recent session of the Committee on forestry, it was reported that the International Panel of Climate Change concluded for North America that disturbances from fire are projected to have increasing impacts on forests and that fires are affecting the carbon pools cycling. While it has to be noted that in many cases forest fires originate from legitimate vital economic interests, careless use of fire in agriculture and pasture lands or for land clearing is causing extended and unintentional damage.

The papers published in this book make an important contribution to our better understanding of forest fires. The Editors hope that the work of the contributors will help to produce recommendations for fire planning and monitoring as well as prevention and rehabilitation.

The Editors are grateful to all contributing authors for the quality of their papers and to the reviewers for helping to select them.

The Editors
Kos, 2010

Contents

Section 1: Computational methods and experiments

Correlation analysis and fuel moisture estimation based on FMA and FMA$^+$ fire danger indices in a *Pinus elliottii* plantation in southern Brazil
J. F. Pereira, A. C. Batista & R. V. Soares 3

Correlations between heat release rate and gaseous by-product concentrations applied to the characterization of forest fuels
I. Fernández-Gómez, J. Madrigal, A. J. de Castro, M. Guijarro, J. M. Aranda, C. Diez, C. Hernando & F. López 15

A comparative study of two alternative wildfire models, with applications to WSN topology control
G. Koutitas, N. Pavlidou & L. Jankovic 25

Diffusion limited propagation of burning fronts
M. Conti & U. M. B. Marconi 37

Statistical parameter estimation for a cellular automata wildfire model based on satellite observations
E. Couce & W. Knorr 47

Sand on fire: an interactive tangible 3D platform for the modeling and management of wildfires
S. Guerin & F. Carrera 57

Section 2: Air quality and health risk models

Numerical modelling of 2003 summer forest fire impacts on air quality over Portugal
A. I. Miranda, V. Martins, M. Schaap, R. San José, J. L. Perez, A. Monteiro, C. Borrego & E. Sá 71

Monitoring fire-fighters' smoke exposure and related health effects during Gestosa experimental fires
A. I. Miranda, V. Martins, P. Cascão, J. H. Amorim, J. Valente, R. Tavares, O. Tchepel, C. Borrego, C. R. Cordeiro, A. J. Ferreira, D. X. Viegas, L. M. Ribeiro & L. P. Pita 83

Section 3: Detection, monitoring and response systems

An integrated approach for early forest fire detection and verification using optical smoke, gas and microwave sensors
N. von Wahl, S. Heinen, H. Essen, W. Kruell, R. Tobera & I. Willms 97

Assessing burn severity using satellite time series
S. Veraverbeke, S. Lhermitte, W. Verstraeten & R. Goossens 107

Real time fire front monitoring through smoke with bi-spectral infrared imaging
J. M. Aranda, J. Meléndez, L. Chávarri & F. López 119

Forestwatch® wildfire smoke detection system: lessons learned from its two-year operational trial
M. Lalkovič & J. Pajtíková 131

Semi-expendable Unmanned Aerial Vehicle for forest fire suppression
D. Benavente 143

Meteorological condition and numerical simulation of the atmospheric transport of pollution emitted by vegetation fires
A. M. Ramos, F. C. Conde, S. Freitas, K. Longo, A. M. Silva, D. S. Moreira, P. S. Lucio & A. L. Fazenda 149

Section 4: Decision support systems

SIRIO high performance decision support system for wildfire fighting in alpine regions: an integrated system for risk forecasting and monitoring
L. Corgnati, A. Losso & G. Perona 163

Innovative image geo-referencing tool for decision support in wildfire fighting
A. Losso, L. Corgnati & G. Perona 173

Section 5: Resources optimization

Allocation of initial attack resources
D. B. Rideout, Y. Wei & A. Kirsch 187

Optimal timing of wildfire prevention education
D. T. Butry, J. P. Prestemon & K. L. Abt 197

Comparing environmental values across major U.S. national parks
D. B. Rideout, P. S. Ziesler & Y. Wei 207

Section 6: Risk and vulnerability assessment

A volatile organic compounds flammability approach for accelerating forest fires
L. Courty, K. Chetehouna, J. P. Garo & D. X. Viegas 221

Forest fires, risk and control
H. Azari 233

Spatial distribution of human-caused forest fires in Galicia (NW Spain)
M. L. Chas-Amil, J. Touza & J. P. Prestemon 247

Evaluation of the FCCS crown fire potential equations in Aleppo pine (*Pinus halepensis* Mill.) stands in Greece
M. D. Schreuder, M. D. Schaaf & Da. V. Sandberg 259

Author Index 271

Section 1
Computational methods and experiments

Correlation analysis and fuel moisture estimation based on FMA and FMA$^+$ fire danger indices in a *Pinus elliottii* plantation in southern Brazil

J. F. Pereira, A. C. Batista & R. V. Soares
Forest Fire Laboratory, Federal University of Paraná, Brazil

Abstract

This research was carried out in a *Pinus elliottii* plantation, established in 1984, with 47.16 $m^2.ha^{-1}$ of basal area, located in the Rio Negro Forest Research Station, owned by the Paraná Federal University, Paraná State, southern Brazil. The research objectives were to analyze the correlations between the FMA and FMA$^+$ fire danger indices and the fine fuel moisture, and develop mathematical models to estimate the fuel moisture based on those indices. The meteorological variables were obtained from the SIMEPAR weather station, located 50km away, and from a pluviograph and a thermo-hygrograph installed in the study area. The dead forest fuels were collected from 30x30cm plots, between 12 noon and 2:00PM, and classified as: AA – surface layer; AB – intermediate layer; AC – lower layer; and B – woody material with 0.7 to 2.5cm diameter. The average fuel layer thickness ranged from 14.8 to 15.3cm. The total fuel load varied from 3185.50 to 4266.01$g.m^{-2}$. The fire danger indices were calculated daily and the values obtained on the fuel collecting days were used to calculate the correlations. The correlation coefficients between relative humidity and fuel classes were 0.42, 0.36, 0.32, and 0.41 for the AA, AB, AC, and B classes, respectively. The correlation coefficients between precipitation and fuel classes were 0.57, 0.38, 0.34, and 0.15 for the AA, AB, AC, and B classes, respectively. Higher correlation coefficients were obtained between fuel moisture and fire danger indices. The correlation coefficients between the fuel classes and the FMA$^+$ were -0.53, -0.56, -0.63, and 0.81 for the classes B, AB, AA, and AC,

WIT Transactions on Ecology and the Environment, Vol 137, © 2010 WIT Press
www.witpress.com, ISSN 1743-3541 (on-line)
doi:10.2495/FIVA100011

respectively. The FMA^+ was the most efficient variable in the modeling development to estimate dead fuel moisture.
Keywords: Pinus elliottii, fire danger indices, forest fuel, forest protection.

1 Introduction

Pinus sp plantations represent approximately 35% of the Brazilian afforested areas, and the State of Paraná, with 37% of the total, ranks first in relation to the total area planted with pine in Brazil [1]. The crescent expansion of the afforested areas, mainly in the country's southern region, requires a continuous improvement in management and protection techniques. Forest fires are a constant threat to the plantations and represent one of the main objectives of the protection plans.

Fuel moisture knowledge is essential to estimate some fire behavior parameters, such as fire intensity and rate of spread, and is an important factor in prescribing a successful controlled burning. Fuel moisture is also important to appraise the forest fire danger [2, 3].

In Brazil the fuel moisture estimation has been done through direct field sampling and laboratory processing. The field samples are weighed (humid mass) and taken to the laboratory to dry until they have reached constant weight, and then, weighed again (dry mass). The relationship between humid and dry masses gives the moisture content of the collected sample.

According to Batista [4], indirect methods could facilitate the fuel moisture content determination, making the work of the technicians responsible for forest protection activities easier. Therefore, correlation analysis between fire danger indices (FMA and FMA^+) and fuel moisture content could become an important tool in forest prevention and suppression actions.

The objectives of this work were to analyze the correlations between the FMA and FMA^+ fire danger indices and the fine forest fuel moisture, and to develop mathematical models to estimate the fuel moisture based on those indices.

2 Methods

2.1 Location

The research was developed in the Rio Negro Experimental Station (Figure 1), owned by the Federal University of Paraná and administrated by the Forest Science Department, located in the south of the Paraná State, approximately 26°04' S latitude and 49°45'W longitude. The mean altitude of the area is 793m above sea level, the annual precipitation is around 1,400mm, and the climate is Cfb, according Koppen classification, with the mean temperature of the hottest month below 22°C, no dry season (driest month with precipitation > 60mm), and more than 10 frosts a year.

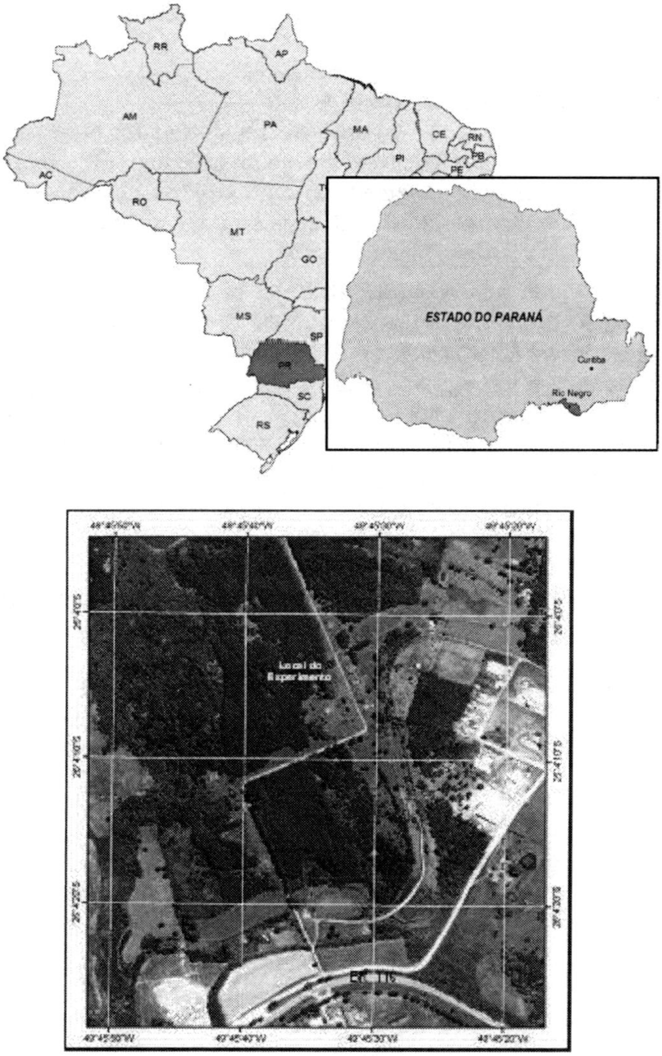

Figure 1: Study area location.

2.2 Meteorological data

According to Brown and Davis [5], the main variables to control the fuel moisture are precipitation, relative humidity, and air temperature. Wind and solar radiation are also important but act indirectly through the fuel temperature modification, the air temperature, and the fuel adjacent thin air layer.

The meteorological data used in this study were obtained from a meteorological station that belongs to the Paraná State official network (Paraná Meteorological System – SIMEPAR), located approximately 50 km from the

Research Station, and also from a pluviograph and a thermo-hygrograph installed in the experiment area.

2.2.1 Sampling and statistical analysis

Site selection for installing the experiment was done after observing the whole stand, looking for a representative area and avoiding the edges. Sampling collection was programmed to extend for a full year, to include the four seasons. It started in the winter of 2007 and ended in the autumn of 2008, always sampling in the driest periods of each season.

The experimental area totalized 200 m^2, divided in four sampling strips, corresponding to the year seasons (Figure 2). The strips were located between the trees lines, in the north-south direction, and the material was collected in the spaces between the trees. Eight samples were collect per day, always between 12 noon and 2:00 PM, during 60 days (15 days per season), totalizing 480 samples. The sampling units measure 30x30 cm. The collected fuel was classified according to Brown et al. [6], using a diameter gauge.

The collected material was divided in two classes: A – needles and small branches with diameter < 0.7 cm; and B – woody material (small branches in different decomposition stages) with diameter between 0.71 and 2.5 cm. Woody material with diameter > 2.5 cm was not sampled due to the high variability and because they take much time to change the moisture content (high timelag).

According to Molchanov [7] the duff layer (decaying leaves and small branches) gets a special structure due to the influence of precipitation, air temperature, cryptogrammic flora, and insects, forming three different strata. Therefore, the A class was subdivided into three sub-classes (Figure 3), as follows:

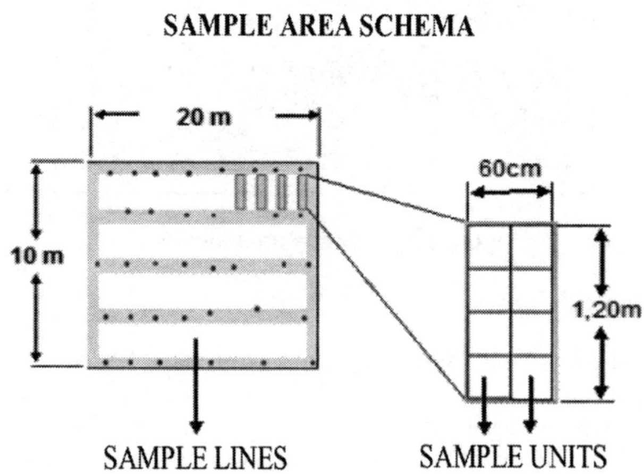

Figure 2: Sampling strips and units.

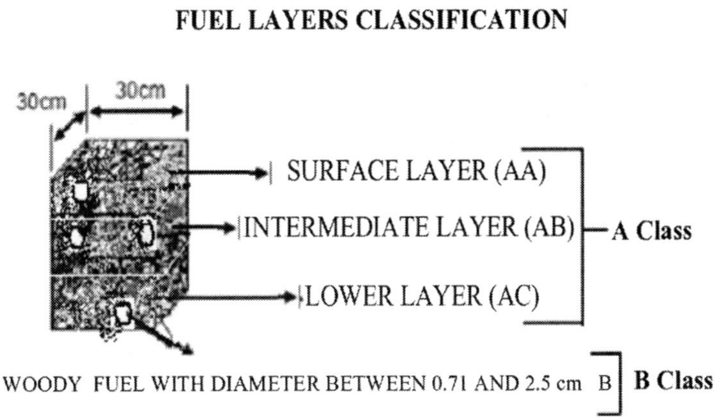

Figure 3: Characterization of the fuel layer, showing the B class and the A sub-classes.

- Surface layer (AA) – composed of needles and small branches (diameter > 0.7 cm); characterized by needles of clear brown color, bright, quite rigid, recently felled.
- Intermediate layer (AB) – also composed of needles and small branches but the needles presented a brown color, bright less, less rigid, indicating the decomposition process beginning.
- Lower layer (AC) – also composed of needles and small branches but the needles presented a dark brown color, low rigidness, and advanced decomposition process.

In the laboratory, the collected fuel was transferred to paper bags and placed in an oven to dry, at 75°C, during 72 hours. After that, the moisture content was determined through the following equation:

$$MC = \frac{Pu-Ps}{Ps}100$$

where:

MC = moisture content in %;
Pu = fuel humid weigh (Just after collected in the field);
Ps = fuel dry weight (after oven dried).

Initially, a correlation analysis including all variables was performed. The mathematical models used to estimate the fuels moisture content were obtained through the backward process, which uses the variables selected by the correlation analysis. To select the best models, two comparing tests were used:

a) Determination coefficient (R^2) – parameter that expresses how much of the dependent variable is explained by the independent variables.

b) Estimation standard error (S_{yx}) – that expresses how much, in average, the observed values varies in relation to the estimated values.

To interpret the variables included in the fuels moisture content mathematical models and in the correlation matrices, the following conventions were adopted (Table 1).

To develop the equations for estimating the fuel moisture content numeric values were used to identify the seasons of the year, as follows:

a) 1 – winter
b) 2 – spring
c) 3 – summer
d) 4 – autumn

The forest fire danger indices (FMA and FMA$^+$) were calculated through a Pascal language program [8], according to the equations developed by Soares [9] and Nunes [8]:

$$FMA = \sum 100 \frac{1}{UR}$$

Table 1: Description of the variables used in the correlation analysis and the mathematical models.

Variables initials	Variables description	Units
E	Season of the year	1 to 4
U_{AA}	Moisture content of the surface fuel (AA)	%
U_{AB}	Moisture content of the intermediate fuel (AB)	%
U_{AC}	Moisture content of the lower fuel (AC)	%
U_B	Moisture content of the woody fuel (B)	%
U_{FZ}	Relative humidity at 1;00PM	%
W_{Lp}	Wind speed (SIMEPAR weather station)	$M.s^{-1}$
P_{FZ}	Precipitation	mm
F_{Fz}	Monte Alegre formula (FMA)	Value
GF_{Fz}	FMA danger degree	1 to 5
F^+_{Fz}	Enhanced Monte Alegre formula (FMA$^+$)	Value
GF^+_{Fz}	FMA$^+$ danger degree	1 to 5

where:

FMA = Monte Alegre Formula
UR = Relative humidity at 1:00 PM

$$FMA^{+} = \sum 100\left(\frac{1}{UR}\right)^{v}$$

where:

FMA^{+}= Enhanced Monte Alegre formula
UR = Relative humidity at 1:00 PM
v = wind speed in m.s^{-1} at 1:00 PM

For the correlation analysis and the fuel moisture content estimation mathematical models, the indices were included according to the daily values and the danger degree scale (Table 2).

3 Results and discussion

The matrix presented in table 3 shows the correlation coefficients among the fire danger indices, the meteorological variables, and the fuels moisture contents. The danger degree levels (1 to 5) presented better results when compared to the daily indices values. The enhanced Monte Alegre Formula (FMA^{+}) presented higher correlation with the fuel moisture than the original FMA, demonstrating that the wind speed inclusion in the original equation improved its performance regarding the correlation with the fuel moisture. It can be observed in table 3 that the correlation coefficients between the danger degree of the FMA^{+} and the fuel moisture of classes AA, AB, AC, and B were -0.63, -0.56, -0.81 and -0.53, respectively, whereas for the FMA the coefficients were -0.60, -0.47, -0.71 and -0.36. The coefficients were negatives because as higher is the fire danger, lower is the fuel moisture. The AC class presented higher association ($r = -0.81$), probably because it is not subject to fast moisture loss or gain, due to its position in the fuel layer.

Table 2: Fire danger degrees of FMA and FMA^{+} used in the correlation analysis and the fuel moisture mathematical models.

Danger degree	Numeric value
Null	1
Small	2
Medium	3
High	4
Very high	5

Table 3: Correlation matrix of the fuel classes moisture in function of the meteorological variables and the fire danger indices.

	U_{AA}	U_{AB}	U_{AC}	U_B	U_{Fz}	W_{Lp}	P_{Fz}	F_{Fz}	GF_{FZ}	F^{+}_{Fz}
U_{AA}										
U_{AB}	0.67**									
U_{AC}	0.57**	0.59**								
U_B	0.61**	0.68**	0.65**							
U_{Fz}	0.42*	0.36*	0.32*	0.41*						
W_{Lp}	0.04	0.16	-0.19	-0.06	-0.17					
P_{Fz}	0.57**	0.38*	0.34*	0.15	0.19	0.28				
F_{Fz}	-0.46**	-0.46**	-0.80**	-0.49**	-0.34*	0.28	-0.34*			
GF_{FZ}	-0.60**	-0.47**	-0.71**	-0.36*	-0.25	0.05	-0.71**	0.74**		
F^{+}_{FZ}	-0.46**	-0.46**	-0.80**	-0.49**	-0.34*	0.28	-0.34*	1.00	0.74**	
GF^{+}_{FZ}	-0.63**	-0.56**	-0.81**	-0.53**	-0.32*	0.12	-0.57**	0.74**	0.93**	0.86**

The B class presented lower association with the FMA^+, perhaps because the timelag depends on the fuel layer thickness or the size of the fuel particles. According to Fosberg and Deeming [10], fuel particles with 0.7 to 2.5 cm diameter (class B) present a 10 hour timelag in the average, against 1 hour average for fuel particles smaller than 0.7 cm.

The correlation coefficients between the fire danger indices (FMA and FMA^+) and the fuel moisture presented better results when compared to the coefficients obtained between the fuel moisture and the meteorological variables, namely relative humidity and precipitation (Table 3).

The models used to estimate the fuels moisture contents, presented in Table 4, were based in the meteorological variables (relative humidity, precipitation, and wind speed) and the fire danger indices (FMA and FMA^+). In the winter, the best estimation was observed in the AA fuel class ($R^2 = 0.59$), using a single

Table 4: Selected mathematical models to estimate fuel moisture content in a pine plantation, in the Rio Negro Experimental Forest, Paraná, Brazil.

Season	Fuel class	R^2	Model
Winter	AA	0.59	$U_A = 311.672 - 55.309\ GF+_{Fz}$
	B	0.28	$U_B = 88.875 + 1.609\ U_{Fz}$
Spring	AA	0.64	$U_A = -41.0643 + 1.5111\ U_{Fz} + 1.7216\ P_{Fz}$
	AB	0.76	$U_{AB} = 184.5024 + 1.4545\ U_{Fz} - 8.0303\ F+_{Fz}$
	AC	0.79	$U_{AC} = 250.5632 + 0.9582\ U_{Fz} - 4.8418\ F+_{Fz}$
	B	0.33	$U_B = 230.3618 - 19.2836\ GF+_{Fz}$
Summer	AA	0.85	$U_A = 97.79 + 0.69\ U_{Fz} + 2.83\ P_{Fz} - 25.55\ GF+_{Fz}$
	AB	0.73	$U_{AB} = 227.14 - 10.41\ P_{Fz} + 33.54\ GF+_{Fz}$
	AC	0.66	$U_{AC} = 335.64 - 32.10\ GF+_{Fz}$
	B	0.60	$U_B = 311.871 - 3.155\ P_{Lp} - 34.019\ GF+_{Fz}$
Autumn	AA	0.82	$U_A = 87.94 + 0.04\ U_{Fz} + 1.40\ P_{Fz} - 1.25\ F+_{Fz} + 2.04\ GF+_{Fz}$
	AB	0.85	$U_{AB} = 293.11 + -0.30\ U_{Fz} - 0.59\ P_{Fz} - 3.03\ F+_{Fz}$
	AC	0.83	$U_{AC} = 424.66 - 5.85\ P_{Fz} - 46.86\ GF+_{Fz}$
	B	0.82	$U_B = 185.62 + 1.18\ U_{Fz} - 3.66\ P_{Fz} - 3.31\ F+_{Fz} + 10.57\ GF+_{Fz}$

variable (GF^{+}_{FZ}). For the B fuel class the best fit was very poor ($R^2 = 0.28$), and for the fuel classes AB and AC, none of the tested models presented reliable estimations.

In the spring, the best estimations were observed for the AB and AC fuel classes, with R^2 equal to 0.76 and 0.79, respectively. In both cases the variables included in the models were U and F^+. The best model for the AA fuel class was obtained through the U and P variables ($R^2 = 0.64$). For the B class the best fit was obtained with the GF variable, but as observed in the winter, the association was very poor ($R^2 = 0.33$).

In the summer, the models selected to estimate the moisture content of all fuel classes presented good fits, with determination coefficients ranging from 0.60 to 0.85. It must be emphasized that the variable GF^+ was selected to compose all the models.

For the autumn, the selected models presented the highest determination coefficients, $R^2 = 0.82$, 0.83, 0.85, and 0.82 for the fuel classes AA, AB, AC, and B, respectively. Generally, the independent variable fire danger index FMA^+ presented better estimations for most of fuel classes, especially in the autumn.

4 Conclusions

The fire danger indices, FMA and FMA^+, presented higher correlation coefficients with the fuel moisture than the isolated meteorological variables.

The FMA^+ presented better results than the FMA and was the most important variable in the fuel moisture content estimation.

Significant meteorological differences between the seasons were observed; therefore, the models developed for each season presented better fits.

The fuels inside the stands presented high moisture content, even when the indices indicated high and very high fire danger.

The use of the FMA^+ to estimate the fuels moisture content produced fast, efficient, and reliable information.

References

[1] Longhi, S. J. *A estrutura de uma floresta natural de Araucaria angustifolia (Bert.) O. Ktze., no sul do Brasil.* Curitiba 1980. Dissertação (Mestrado em Engenharia Florestal) – Universidade Federal do Paraná.

[2] Fosberg, M. A., Lancaster, J. W. & Schroeder, M. J. Fuel moisture response – Drying relationships under standard and field conditions. *Forest Science*, Lawrence, v. 16, p. 121-128, 1970.

[3] Yebra, M., Chuvieco, E. & Riaño, D. Investigation of a method to estimate live fuel moisture content from satellite measurements in fire risk assessment. *Forest Ecology and Management*, Amsterdam, v. 234, Supl. 1, p. S32, 2006.

[4] Batista, A. C. *Determinação de umidade do material combustível sob povoamentos de Pinus taeda L. no norte do Paraná.* Curitiba, 1984. 61p.

Tese (Mestrado em Engenharia Florestal) - Setor de Ciências Agrárias, Universidade Federal do Paraná.

[5] Brown, A.A. &, Davis, K.P. *Forest fire: control and use*. 2. Ed. New York: McGraw Hill Book, 1973. 686p.

[6] Brown, J. K., Oberheu, R. D. & Johnston, C. M. Handbook for inventorying surface fuels and biomass in the Interior West. Ogden, *Intermountain Forest and Range Experiment Station*, 1982. 48p. (General Technical Report INT-129).

[7] Molchanov, A. A. *Hidrologia Florestal*. Fundação Calouste Gulbenkian. Lisboa, 1965. 419p.

[8] Nunes, J. R. S. FMA, *Um Novo Índice de Perigo de Incêndios Florestais para o Estado do Paraná, Brasil*. Curitiba, 2005. Tese (Doutorado em Engenharia Florestal) – Setor de Ciências Agrárias, Universidade Federal do Paraná.

[9] Soares, R. V. *Determinação de um índice de perigo de incêndio para a região centro-paranaense, Brasil*. Turrialba, Costa Rica, 1972. Tese (M.Sc. en Ciencias Forestales), Centro Tropical de Enseñanza y Investigación, Instituto Interamericano de Ciencias Agrícolas de la OEA.

[10] Fosberg, M.A; Deeming, J.E. Derivation of the 1 and 10 – Hour timelag fuel moisture calculations for fire – danger rating. U.S.D.A. *For service; Research Note RM* – 207, 1971. 8 p.

Correlations between heat release rate and gaseous by-product concentrations applied to the characterization of forest fuels

I. Fernández-Gómez[1], J. Madrigal[2], A. J. de Castro[1], M. Guijarro[2], J. M. Aranda[1], C. Diez[2], C. Hernando[2] & F. López[1]
[1]*LIR laboratory, Departamento de Física, Universidad Carlos III de Madrid, Spain*
[2]*Centro de Investigación Forestal, Instituto Nacional de Investigación y Tecnología Agraria y Alimentaria (CIFOR-INIA), Spain*

Abstract

In this work an adapted bench-scale Mass Loss Calorimeter (MLC) device is used to measure HRR for forest fuels. The MLC has the same heating unit as a standard cone calorimeter, but a) the physical basis to measure HRR in a MLC (by using a calibrated thermopile) is different than the one used in the standard cone calorimeter (oxygen consumption method) and b) the MCL does not have a unit to measure the concentration of the gases produced during the combustion. Although the concentration values are not essential to measure the HRR curves, their knowledge is of great interest to characterize the combustion process and the combustion efficiency. In this sense, a Fourier transform based spectroradiometer (FTIR) has been adapted to the MLC in a short open-path configuration to measure "in situ" the concentration of carbon monoxide and dioxide and water vapour, nearly simultaneous to the measurement of the HRR values. This simultaneity in both types of measurements allows one to find correlations between different variables. These correlations would help to make predictions on unknown variables in the framework of fire models.
Keywords: calorimetry, heat release rate, forest fuels, short open path FTIR spectroscopy.

WIT Transactions on Ecology and the Environment, Vol 137, © 2010 WIT Press
www.witpress.com, ISSN 1743-3541 (on-line)
doi:10.2495/FIVA100021

1 Introduction

The characterization of combustion properties and flammability of forest fuels is not a straightforward task. Forest fuel combustion is a complex process with multiple interrelated components, some of which have not yet been measured. There is a general agreement that the Heat Release Rate (HRR) of a fuel is one of the most important characteristics for understanding the combustion process, fire characteristics and fire propagation rate. Physical models take into account the complex phenomena to relate combustion variables (energy and gases emitted) with fire front behaviour. Nevertheless, validation of the prediction offered by models is complex because instrumental techniques are not available to measure HRR and gases directly. In addition, understanding the complex forest fire combustion necessarily involves the simulation of the phenomena at the bench-scale approach. However, there is no universally accepted methodology for forest fuels, and many approaches have been evaluated for applying bench-scale devices to the study of these types of fuels.

The quantification of the frontal fire intensity of fires, expressed as heat-release rate per unit length is usually estimated from the mass loss rate through the Byram equation:

$$I = H \cdot w \cdot r$$

where I is the frontal fire line intensity (kW/m), H is the heat of combustion (kJ/kg), w is the fuel consumption on an area basis (kg/m^2) and r is the fire spread rate (m/s). There is a controversy about the correct value of H used for forest fuels. Several authors propose the use of the net heat of combustion obtained in an oxygen bomb, using 18 MJ/kg as a medium value for forest fuels. Nevertheless other authors proposed a value of 15 MJ/kg, incorporating a nominal 15% radiation loss and an additional heat loss due to evaporation of all fuel moisture. This value is the upper limit obtained for flaming combustion of conifers in large scale experiments (12-15 MJ/kg), showing the importance of determining the heat of combustion during the flaming phase, which is much lower than in the glowing phase and strongly dependent on moisture content. Forest modellers traditionally do not pay attention to this variable because it is considered that it introduces little error into the energy calculation compared with r and w, and because it has been considered as a constant. Calorimetry studies show the significant differences of HRR and H among species, so the influence of these variables during the forest fire behaviour must be clarified. On the other hand, bulk density has important implications in flammability because forest fuels are irregular porous fuels and the natural diffusion of air affects the combustion process. To sum up, the complexity of the heat release estimation in forest fires is limited by the correct measure of variables involved. The need to understand the complex forest fire combustion (rapid flaming combustion in porous fuel with a low bulk density along a dynamic fire front) necessarily involves the simulation of the phenomena at the bench-scale approach.

In this work a specific study has been performed that is focused on studying the influence of the fuel moisture content and bulk density on the measurements performed by the MLC-FTIR.

2 Experimental

2.1 Experimental devices

2.1.1 The Mass Loss Calorimeter (MLC)

The Mass Loss Calorimeter (MLC) was manufactured by Fire Testing Technology Limited (FTT®). This apparatus (fig.1) is the complete fire model of the cone calorimeter, which has assumed a dominant role in bench-scale fire testing of building materials. A chimney made of stainless steel (600 mm long x 114 mm inner diameter) and containing a thermopile of four mineral insulated inconel sheathed thermocouples (type K, 1.6 mm diameter) was added to the MLC (650 mm above the holder surface). The thermopile output was first calibrated by use of a methane burner and a flow meter, and then used to quantify heat release [1].

The MLC standard sample holder contained low density ceramic wool to ensure correct positioning of the samples, 25 mm from the conical heater, and the sample was placed on aluminium foil. A specific holder adapted for forest fuels samples was also designed to simulate rapid flaming combustion [2]. The holder (10 x 10 x 5 cm^3) was made of stainless steel, with small uniformly sized holes over the entire outer surface (sides and bottom). These holes create an open space for inlet combustion gases to pass into the holder and through the fuel samples (Figure 1).

The MLC device and the porous holder have been evaluated and comply with the repeatability criteria established by different authors [3-5].

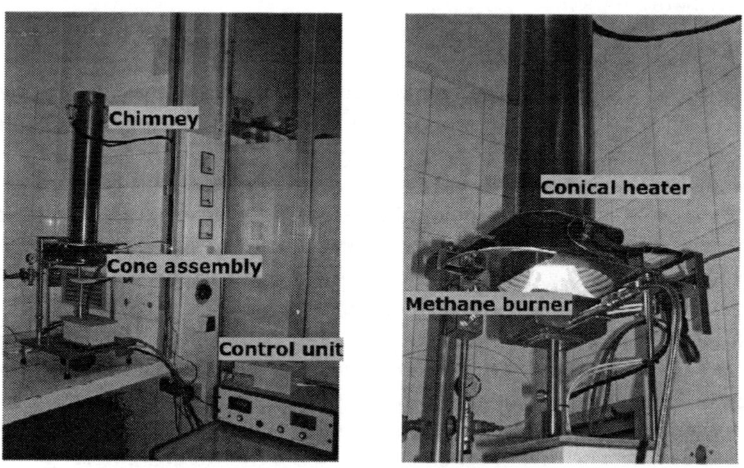

Figure 1: Experimental device at the INIA-CIFOR laboratory. (Left) General view. (Right) Methane burner calibrating the thermopile.

2.1.2 The open-path FTIR spectroradiometer

Traditionally, extractive methods are the most common ones to analyze gas composition in a great variety of problems. However, they present some issues that must be considered carefully. The most important is the need to conduct the gas sample to the analyzer, a process that can involve modifications in the chemical composition of the analyzed gas.

Remote sensing techniques overcome some of these problems. One of the main advantages of remote sensing is that it is a non intrusive method that does not require the collection of samples, avoiding any alteration of the analyzed gas. In this sense, open-path FTIR Spectroscopy appears to be a very interesting technique that combines the advantages of the FTIR spectroscopy with the remote sensing principles. The open-path setup consists of a source of infrared energy and a FTIR spectroradiometer that measures the infrared energy coming to the instrument. The analysis of the absorption bands provides qualitative and quantitative information on the gases that are present at the path between the IR source and the spectroradiometer.

In this work an FTIR spectroradiometer working in an open-path configuration has been coupled to the MLC to analyze in-situ gas concentrations. In this sense, the MLC appears to be the most interesting configuration to do that. The heat unit in the MLC is exactly the same than in a standard cone calorimeter. Instead of the complex exhaust and gas sampling and analyzing system, the MLC has a "chimney-like" thermopile. The main function of the thermopile is to measure the heat release rate curves, but for our purposes also can serve as a duct to conduct the gaseous by-products of the combustion. Then the open-path system can be mounted in such a way that the optical line of sight of the spectroradiometer is only a few cm above the exhaust duct. In this way, radiation coming from the hot metallic wall of the thermopile is avoided, and only absorption from the gases at the exhaust will be measured. Fig. 2 shows the proposed configuration.

The main characteristics of the open-path system used for these experiments are:

a) The infrared source is an electric radiator powered at 400 W. In this way, the surface reaches a temperature around 600°C working as a very nice IR radiator in the medium infrared (MIR) spectral range.

b) The spectral resolution selected has been 0.5 cm^{-1} (the best one that provides the MIDAC-AM model of spectroradiometer) in order to measure properly the fine structure of the CO absorption band and to take advantage of this resolution to retrieve in the best experimental conditions other gases. For this resolution, each spectrum takes 1.7 seconds to be acquired.

c) The number of scans selected is two. This is the most adequate value that minimizes the acquisition time preserving and adequate signal-to-noise ratio.

Figure 2 presents a scheme of the typical experimental set up. Distances between the infrared source and the spectroradiometer are around 320 cm. It is important to note that this distance is not critical for the quantitative retrieval of concentrations, although it is very convenient to maintain it for the different experiments in order to assure a similar level of energy impinging at the detector.

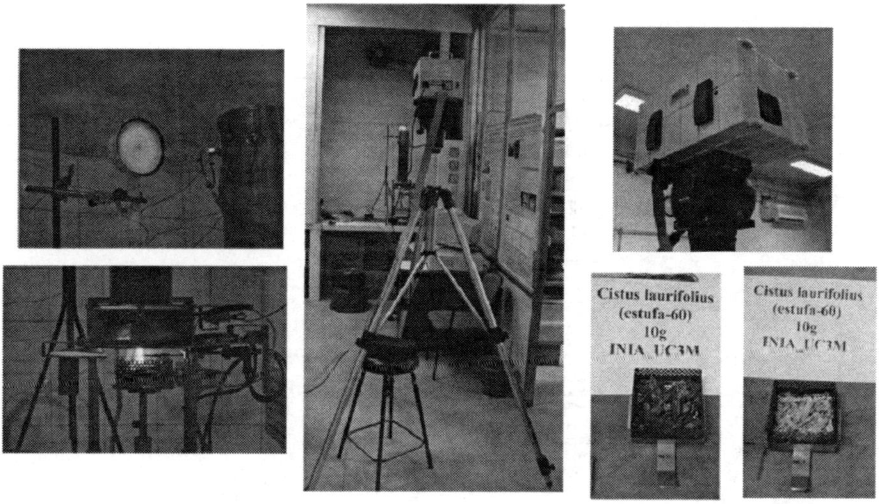

Figure 2: General view of the MLC-FTIR set up (centre) and different details of the experimental configuration.

2.2 Sample preparation

A series of tests, using *Cistus ladanifer* L. samples (leaves and twigs < 1cm diameter) was carried out to determine the combustion characteristics of the forest fuel bed. The fuel moisture content (FMC) was controlled. The resulting FMC's, calculated on an oven-dry basis after drying the samples at 60°C to constant weight, were ~110%, ~75%, ~40% and 0% (oven-dry). Three replicates were tested for each holder in order to comply with the repeatability criteria (n=12). The initial sample dry mass selected was 10 g and the resulting thickness of the mass was 5 cm. In accordance with the volume of the holder, the experimental conditions correspond to a bulk density of ~20 kg/m^3 (representative of a bulk density value under field conditions). A constant heat flux of 50 kW/m^2 was selected in the electric conical heater for exposure of the samples because a similar value was expected in the wind tunnel tests. The MLC adapted design porous holder was used. The sample uniformly covered its exposed surface area. The spark igniter was used to provide the piloted ignition [6].

3 Results and discussion

3.1 Repeatability of the measurements

Fig. 3 illustrates the level of repeatability expected for these experiments. Three replicates have been tested for each experimental condition.

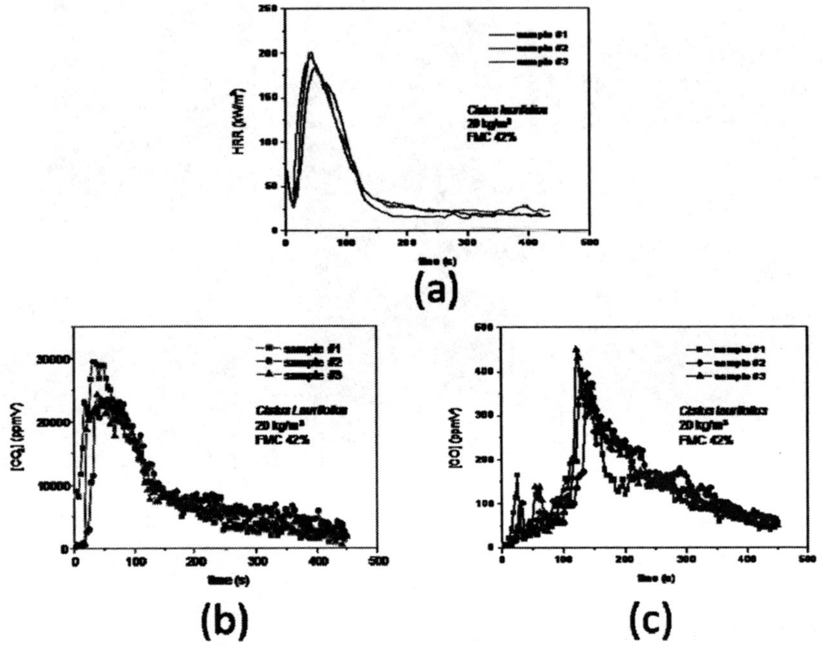

Figure 3: Experimental curves of HRR (a), CO_2 (b) and CO (c) volume mixing ratios measured for a series of three replicates with *C. laurifolius* and a fuel moisture content of 42%.

As can be seen, an acceptable repeatability is obtained. Taking into account the difficulty to work with biomass as a sample, this repeatability is indicative of an appropriate sample preparation procedure.

3.2 Data analysis

Fig. 4 is an example of the correlations between the thermodynamic variables and the emission of gaseous products as a function of time. Two different regimes (flaming and non-flaming) for the combustion can be clearly determined by studying the temporal evolution o these variables. Most of the heat is released during the flaming combustion, which is characterized by a good oxidation of the carbon fuel to a CO_2 gaseous phase. During the smouldering combustion the released heat tends to be negligible, whereas a poorer combustion with predominant oxidation of the carbon to a CO gaseous phase is clearly detected. As can be seen, the temporal evolution of the CO concentration is a very good indicator of the state of the combustion process, and it is easy to identify and separate from this evolution the flaming and the smouldering phases. HRR curves are clearly related to the flaming combustion, when most of the CO_2 is released.

The results are shown as curves of HRR plotted against time (1 second frequency) and the following numerical results from the series of tests: Time to Ignition (TTI, s) , Flame Duration (FD, s), FD before time to peak HRR (bFD, s), Peak HRR (PHRR, kW/m^2), average of HRR during flaming combustion (HRR, kW/m^2), Total Heat Release during flaming combustion (THR, MJ), THR before time to peak HRR (bTHR, MJ), Average Effective Heat of Combustion during flaming phase (AEHC, MJ/kg), Peak Effective Heat of Combustion (pEHC, MJ/kg), Average Effective Heat of Combustion before time to peak EHC (bEHC, MJ/kg), Average Mass Loss Rate during flaming phase (MLR, g/s), peak Mass Loss Rate (pMLR, g/s), MLR before time to peak MLR (bMLR, g/s), Residual Mass Fraction (RMF, %) and Residual Mass Fraction before time to peak HRR (bRMF, %).

An exploratory analysis was developed using non-parametric tests (Spearman R tests) in order to relate peak CO_2 concentration with combustion parameters during flaming phase. FMC was also considered as independent variable in order to detect the influence in maximum CO_2 concentration. The Partial Least Square (PLS) regression model (SIMPLS algorithm) was used to explore the relationship between peak CO_2 (considered as dependent variable) and the most significant combustion parameters previously detected (considered as predictive variables). Statistica 6.0 package® was used to analyze these data.

Fig. 5 shows HRR and [CO_2] curves for different moisture contents tested. The typical progression of a test is shown: ignition is produced, the heat release rate rises quickly and the peak (PHRR) is reached, then the HRR decreases until

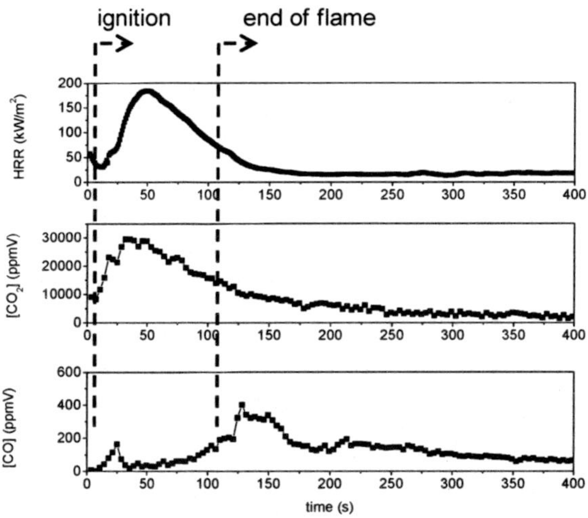

Figure 4: A comparison of the temporal evolution of different magnitudes measured for *C. laurifolius* with a fuel moisture content of 42%. The beginning and end of the piloted flaming combustion is indicated by the dashed lines.

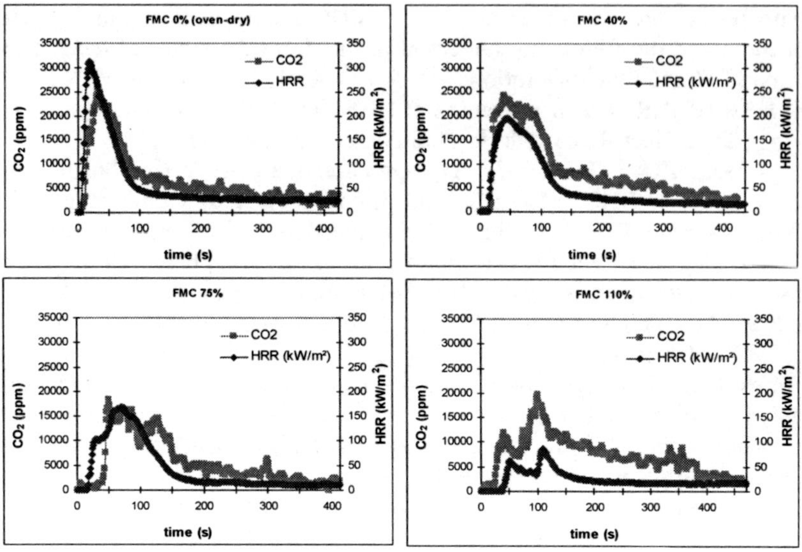

Figure 5: HRR and $[CO_2]$ curves for each FMC tested.

Table 1: Paired-correlations between $[CO_2]$ and independent variables analysed.

	Spearman R	p-level
CO_2 & FMC	-0,693530	0,017943
CO_2 & TTI	-0,633259	0,036475
CO_2 & FD	-0,246014	0,465874
CO_2 & bFD	-0,493156	0,123222
CO_2 & HRR	0,672727	0,023313
CO_2 & pHRR	0,609091	0,046696
CO_2 & bEHC	0,863636	0,000612
CO_2 & pEHC	0,490909	0,125204
CO_2 & AEHC	0,618182	0,042646
CO_2 & bMLR	0,454545	0,160145
CO_2 & pMLR	0,451026	0,163816
CO_2 & MLR	0,045455	0,894427
CO_2 & bTHR	-0,009091	0,978837
CO_2 & THR	0,645455	0,031963
CO_2 & bRMF	-0,290909	0,385457
CO_2 & RMF	0,463636	0,150901

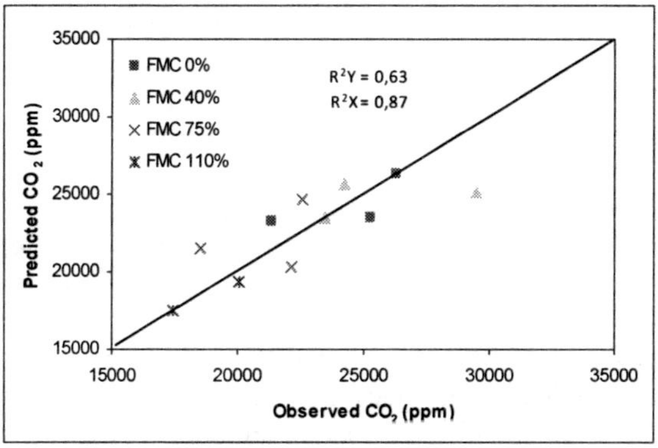

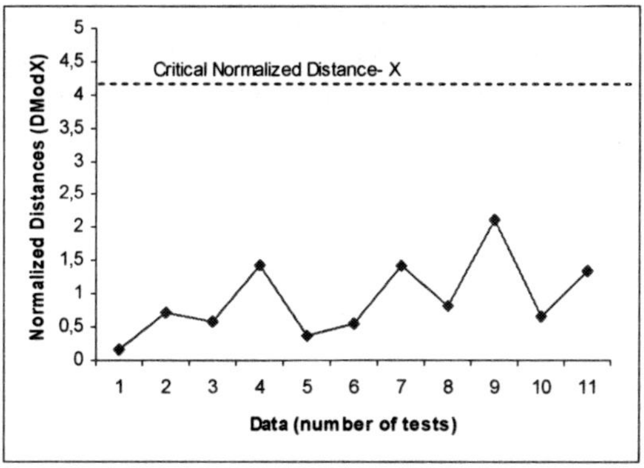

Figure 6: Goodness of fit and residual analysis for the partial least squares (PLS) model developed.

the end of flame and finally the glowing phase starts and continues until the end of the test. Curves also describe that FMC reduce time-to-ignition and HRR [25-26] for the same bulk density (20 kg/m^2)

Paired-correlations between peak of CO_2 concentration (ppm) and selected combustion characteristics have shown in table 1. There was not flaming phase for one of the test for FMC 110% so this test was removed from the correlation analysis (n=11). Results show a significant and positive correlation between peak CO_2 concentration and typical combustion characteristics (MLC output) such us TTI, HRR, pHRR, AEHC and THR. Results also show the positive significant influence of time-heat flux history before the peak HRR (bEHC, kJ/kg) in peak CO_2. The negative significant correlation between FMC and [CO_2] ratify the observed effect of FMC in combustion process detected in HRR curves.

PLS model was developed to relate CO_2 concentrations with combustion characteristics using as predictors the significant variables previously detected. The results show that predictors (FMC, TTI, HRR, pHRR, AEHC, THR and bEHC) explain 63% of the variability of CO_2 concentration (R^2Y=0.63, Second

Component, n = 11). PLS model predicts reasonable well observed values and anomalous residuals were not detected (fig. 6).

4 Conclusions

An FTIR spectroradiometer has been coupled in a short open path configuration to measure in situ concentrations of CO, CO_2 and H_2O obtained as combustion product of forest fuels during test performed in a Mass Loss Calorimeter. The ignition and flame time can be measured using the temporal evolution between the concentrations of CO and CO_2 with the HRR. Correlations between these concentrations and typical magnitudes measured in a calorimetric test have been studied for different values of fuel moisture content in order to obtain prediction capabilities. Results show a significant and positive correlation between peak CO_2 concentration and typical combustion characteristics (MLC output) such us TTI, HRR, pHRR, AEHC and THR and the influence of time – heat flux history before the peak HRR in peak CO_2 concentration.

Acknowledgement

The authors want to acknowledge financial support from the Integrated Project FIRE PARADOX, FP-018505.

References

[1] International Organization for Standardization (2001). Simple heat release test using a conical radiant heater and a thermopile detector (ISO 13927), International Organization of Standardization, Geneva.

[2] Schemel, C.F., Simeoni, A., Biteau, H., Rivera, J.D. & Torero, J.L. A calorimetric study of wildland fuels, *Experimental Thermal and Fluid Science* **32 (7)**: pp. 1381-1389, 2008

[3] Babrauskas, V. The cone calorimeter, in: *SFPE handbook of fire protection engineering,* 3rd ed, National Fire Protection Association, Quincy MA, pp. 3-63 – 3-81, 2002.

[4] European Commission (1997). SBI round robins results Available from http://europa.eu.int/comm/enterprise/construction/internal/essreq/fire/sbiround/sbirep.htm .

[5] Janssens, M.L. Heat Release Rate (HRR), ,in: Measurement Needs for Fire Safety, Proceedings of an International Work-shop (NISTIR 6527), T.J. Ohlemiller, E.L. Johnson and R.G. Gann (Ed.), National Institute of Standard and Technology, Gaithersburg. pp. 186-200, 2001

[6] Madrigal, J,, Hernando, C., Guijarro, M., Diez, C., Marine, E. & de Castro, A.J. *Journal of Fire Sciences* **27**, pp. 323-342, 2009

A comparative study of two alternative wildfire models, with applications to WSN topology control

G. Koutitas[1], N. Pavlidou[1] & L. Jankovic[2]
[1]*Department of Electrical and Computer Engineering, Aristotle University of Thessaloniki, Greece*
[2]*Intesys Ltd, Birmingham Science Park, UK*

Abstract

In this paper two wildfire modelling methods are compared in terms of performance, scaling up flexibility and speed of model execution. The InteSys model is based on Cellular Automata (CA). Simple rules are applied to each cell, interacting with neighbouring cells. The cell based structure reflects the object oriented nature of the model, as each cell is a working copy of a cell class – a blueprint that enables easy expansion, taking into account undergrowth, tree spacing, moisture content, air temperature, solar radiation, wind velocity, terrain gradient, tree flammability, and other parameters. The CD-AUTH model is based on the Cell-DEVS technique operating also on a domain discretized to interacting cells, incorporating the same as above physical properties, variable in time and coupled to a low level surface wind module. The model applies the Rothermel approach with respect to the fire propagation considering the Huygens ellipse of propagation. Advantages and disadvantages of the two models are discussed on the basis of comparative simulations over hypothetical fire scenarios on a digital map. Important observations and conclusions are also drawn concerning the deployment of wireless sensor networks (WSN) for wildfire detection. Finally, a network topology control algorithm that utilizes the fire prediction algorithms is presented and yields energy efficiency of the WSN, providing with high time resolution data for real time monitoring.

Keywords: wild fire modelling techniques, cellular automata, discrete event simulations, cell-DEVS, wireless sensor networks WSNs, network topology control, energy efficiency WSNs.

1 Introduction

Forest fires detection holds an important role in fire management and different detection strategies have been applied to monitor large areas. These can be automatic video surveillance systems, Unmanned Aerial Vehicles (UAV), satellite imagery and wireless sensor networks (WSN) [1]. The estimation of the risk of ignition of a wildfire in forests is the first step to fire management. That risk is quantified according to the fuel available and the weather conditions via the algorithm of the FWI (Fire Weather Index), established in Canada [2]. Wireless sensor networks are considered as a scalable solution that can provide real time fire detection and monitoring of the crucial parameters of FWI, overcoming limitations of the above mentioned alternative detection techniques [1]. In [3–5], various forest fire detection techniques that are based on the WSNs are presented. Furthermore, WSNs can provide real time measurements of critical parameters to the fire propagation algorithms and this can yield accuracy improvements of the models and better fire predictions and management.

An effective strategy to manage wildfires is based on the detection system used and the algorithm implemented to model the fire propagation in the area of investigation. In general, three alternative modelling techniques exist, namely the empirical, semi-empirical and physical [6]. Semi-empirical models are preferred for engineering application since they produce accurate results with low CPU demands. Rothermel [7] first described fire spread as a mathematical model. Software tools and semi-empirical models are now based on the integration of the Rothermel's equation integrated with cellular automata (CA) or discrete event (cell-DEV) approximation to model the fire spread over digital elevation maps and GIS and are considered as the most suitable approximations. Cellular models of fire growth use fixed distances between regularly spaced grid cells to solve the fire arrival time from one cell to another. There are several types of CA models for fire growth, including the transfer of fractional burnt area, probability driven models and fractal models [8-11]. DEVS are applied to define arbitrary ordinary differential equations. A system model of DEVS is described as a hierarchical composition of submodels each of them being behavioural or structural. Cell-DEVS formalism is a combination of DEVS and CA [12, 13].

In this paper two wildfire modelling methods are compared in terms of performance, scaling up flexibility and speed of model execution. The Intesys model is based on CA approximation being probabilistic in nature with low CPU demands whereas the CD-AUTH model is based on cell-DEVS approximation taking into account the main parameters affecting fire spread from Rothermel's equation and it is coupled to a low level surface wind module for increased accuracy. Consequently, this model has higher CPU demands. An algorithm that enables the use of fire predictions models to WSN topology control is also presented. The fire model is used to predict the growth of fire and feedback the network to provide increased FWI sampling at specific locations, necessary for high resolution in time information to fire fighters and fire management. For the purpose of our investigation the CD-AUTH model was utilized.

2 InteSys-model

The InteSys Event Propagation Model is based on cellular automata machines. Simple rules are applied to each cell, with an interaction framework that operates between neighbouring cells. The system model is not explicitly programmed but it emerges from the component models and their interaction. The cells have geographic connotation and correspond to a raster grid of predefined size, with square cells typically between 10m and 100m sides. The cell based structure reflects the object oriented nature of the model, where each cell is a working copy of a cell class – a blueprint that enables easy expansion of model capabilities, taking into account undergrowth, tree spacing, moisture content, air temperature, solar radiation, wind velocity, terrain gradient, tree flammability, and other parameters. The working copies of the cell class are instantiated at the start of the simulation, and private values of variables in each instance are created either from a GIS data input or from a command file. For each cell, the model employs Moore neighbourhood of 8 cells to perform calculations and derive the status of each cell (Figure 1a)).

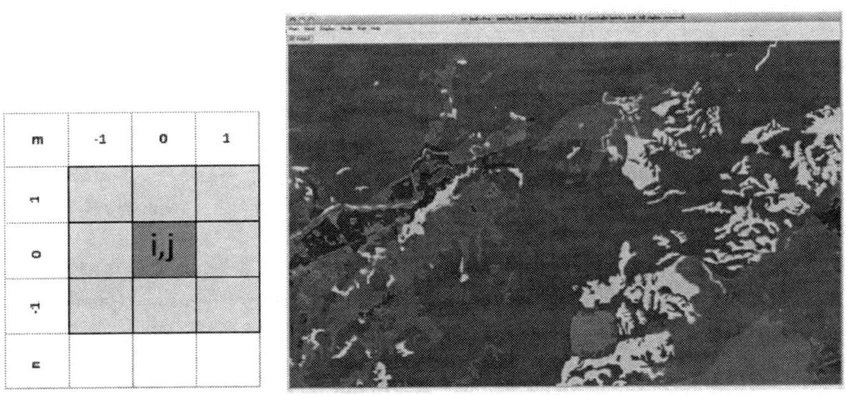

a) b)

Figure 1: a) A land cell in position (i, j) in a Moore neighbourhood of 8 cells, b) IntEvPro in operation: after importing an external GIS file and setting relevant parameters, the model simulates the spread of fire in the forest (dark blue cells) and in open areas (yellow cells). The fire is shown as an expanding circular front in the lower end of the centre of the screen. The simulation time, corresponding to the real time, is shown in the upper left corner.

Wind direction is detected in one of 8 compass directions that correspond to the geometric relationship between the cell and its neighbourhood. For instance, wind from south west comes from the lower left corner of the neighbourhood, from position i-1, j-1. Direction is calculated as d=10*m+n (fig. 1a), which gives 8 unique numbers, avoiding duplication in direction references. Response to wind and slope is calculated using Rothermel's equation (2).

Fire ignition: Cells are ignited either randomly, or manually using the pointing device, or using a built in preset location.

Fire propagation through the cell: As fire propagates differently in different cell types, each time a cell is ignited, a burning counter starts and compares its total with a number that corresponds to the burnt down state of that cell type. The slope and wind coefficients reduce the counter's total and thus modify the rate of fire propagation in the cell.

Fire propagation between cells: Neighbouring cells catch fire from burning cells with a certain probability, representing a resistance of fire transfer from one cell to another. This probability is modified using slope and wind related propagation coefficients.

Figure 1b) shows the model in operation using an external GIS file with cell size of 20 m x 20 m, and representing the total area size of 25.7 x 17.6 km. The GIS map that represents the cell types if the main output screen, whilst the map with cell altitudes is used for background calculation of fire propagation parameters.

3 CD-AUTH model

3.1 Model description

CD-AUTH is based on the Rothermel's equations [7], for the description of the fire physics i.e. the thermal energy balance along the propagating fire front, its generation on a burning area and its distribution to fractions of vertically convected energy, radiated energy and energy consumed for the combustion of the adjacent fuel. In order to tackle the spatiotemporal variability of the fire evolution over a realistic topography, due to variable fuel loads, humidity, ground slope, wind intensity and direction etc, the model follows the formalism and algorithmic structure deriving from the timed Cell-DEVS methods [11, 12]. The fire domain is discretized in square cells (Figure 2a)) characterized by pertinent state parameters. The fire is introduced initially at a pre-determined cell and the evolution over the 2D domain is controlled by transitions processes in each cell and between adjacent cells. In each cell of the considered 'cellular automaton', a discrete event simulation is applied, and the system is composed of a large number of interacting individual cells (following a strict procedure), controlled by time delays. The magnitudes produced by Rothermel's equations, are the rate of fire spread, and the fireline intensity (deducing the transition from ground fire to crown fire). These equations are applied locally as a 1D model over the area of one cell.

The model makes use of the Huygen's principle [11] locally, using the geometry of the elliptically extending fire front, having as focus the cell centre and dimensions of the ellipse depending on the superimposed local wind and ground slope magnitudes (Figure 2b)). That principle is used to convert in a controlled manner from the one dimensional cell domain (a cell over which the main direction and the maximum rate of fire spread is calculated by Rothermel' equations), to the two dimensional topography of the burning wildland.

The model receives as input the individual cell fuel properties, the topographic data for the estimation of the ground slope, and the local wind speed and direction. The fuel properties and the wind data can be varying in time, to incorporate scenarios of rain or fire combating from the air, as well as any change of wind direction and intensity.

From the above data the "effective" fire direction and maximum propagation rates are computed as well as the 2D rate of spread along the 8 main compass directions connecting each cell with the adjacent cells, according to the preferred square grid discretization (dx of Figure 1a)). Each cell is characterised by an index specifying the transition of state between a non burning (index=0), a burning (index=1) and a burnt (index=2) cell.

According to the composed algorithm, during each time step the following checks are done over the fire domain

1. check for any variation of the cell state variables
2. check for the spread of the fire from any burning cell to the neighbouring cells
3. check for the consumption of the available fuel in a burning cell.

Mathematically, the CD-AUTH model is defined as:

$$CD-AUTH=<K,X,S,G,t,I,E> \tag{1}$$

where K is the set of points with coordinates, i, j in the region of interest (Figure 2a)), X is the geometrical pattern of the cells and defines the change in the state of (Figure. 2b)), S is the state of the cells set that incorporates values representing altitude, fuel characteristics, fire duration, wind direction, wind speed, fire spread. G is the set of global variables that affects the transition functions of the cells and incorporates values such as weather conditions, wind direction and speed, fuel apothem of the cell, t is the transition function set for surface and crown fire spread according to fuel apothem and wind characteristics, I is the ignition cell, E is the external function set.

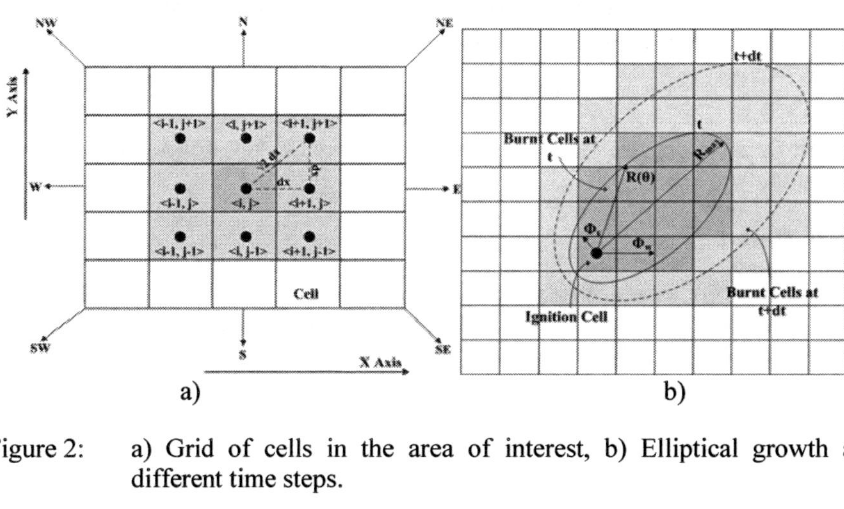

a) b)

Figure 2: a) Grid of cells in the area of interest, b) Elliptical growth at different time steps.

3.2 Surface fire

The fire spread rate is computed according to

$$R = ROS \cdot (1 + \Phi_s + \Phi_w) \tag{2}$$

where

$$ROS = \frac{I_R \cdot \xi}{\rho_b \cdot h \cdot Q_{ig}}$$

$$\Phi_s = a_s \beta^{-b_s} (\tan\phi)^2 \tag{3}$$

$$\Phi_w = C(a_w U)^B (\beta / \beta_{op})^{-E}$$

In the above equations R is the computed rate of spread, I_R is the reaction intensity, ξ is the propagation flux ratio, ρ_b is the ovendry bulk density, h is the effective heating number, Q_{ig} is the heat of preignition, β is the packing ratio, β_{op} is the optimum packing ratio, φ is the terrain slope [14]. Φ_s, Φ_w represent the terrain slope and wind effects to rate of spread. The parameters incorporated in these equations can be found in [7]. The combined terrain and wind effects are computed according to $\vec{\Phi} = \vec{\Phi}_s + \vec{\Phi}_w$

The fireline intensity is computed according to

$$I_b = q \cdot w \cdot R \tag{4}$$

where q represents the net heat produced and w the weight of the fuel per unit area burned in the flaming front [9]. In an arbitrary direction, the spread rate is computed according to an elliptical model, similar to Huygens approximation, and the fire origin is assumed to be on one of the foci (Figure 2b)) according to $R(\theta) = R \cdot (1 - \varepsilon)/(1 - \varepsilon \cdot \cos\theta)$ and the eccentricity of the ellipse is given by $\varepsilon = \sqrt{l_w^{\,2} - 1} / l_w$. Parameter l_w is the semi-major over the semi-minor ellipse ratio and depends on the effective midflame windspeed U_{eff} that considers the wind and slope effects according to (3). It is given by

$$l_w = 1 + l_x (e^{a_x U_{eff}} - 1) + l_y (e^{-a_y U_{eff}} - 1) \tag{5}$$

where l_x, l_y, a_y, a_x are constant values obtained by the Anderson's empirical formulations [14].

3.3 Crown fire

The crown fire effect becomes important if the surface fireline intensity I_b presented in (4) is greater than a threshold value I_0 [11, 14]. The crown fire spread rate is computed according to (6). Parameters c_c and d_c are constant with time [14].

$$R_c(\theta) = R(\theta)\left[1 + c_c\left(1 - e^{-d_c \frac{I_b - I_0}{I_b} R(\theta)}\right)\right] \tag{6}$$

3.4 Low level surface wind module

The wind over an irregular terrain is affected by the obstructions imposed by the hills and mountains of the scenario. In most cases, the input parameters to (1) concerning the wind speed and direction are extracted by sparse meteorological stations or are assumed homogeneous in all the investigated scenarios. In the CD-AUTH model a deterministic low level wind model (LLWM) is coupled to provide a high resolution wind characteristic at each cell. A numerical solution by an explicit centered first order finite difference scheme on the staggered grid (Figure 2a)) was used. The LLWM is defined by the set of equations

$$\begin{aligned} \frac{Du}{Dt} &= -g\frac{\partial \zeta}{\partial \mathrm{x}} + N\nabla^2 u - \frac{C_b}{h}u\sqrt{u^2+v^2} \\ \frac{Dv}{Dt} &= -g\frac{\partial \zeta}{\partial \mathrm{y}} + N\nabla^2 v - \frac{C_b}{h}v\sqrt{u^2+v^2} \\ &\frac{\partial \zeta}{\partial \mathrm{t}} + \frac{\partial h}{\partial \mathrm{x}} + \frac{\partial h}{\partial \mathrm{y}} = 0 \end{aligned} \tag{7}$$

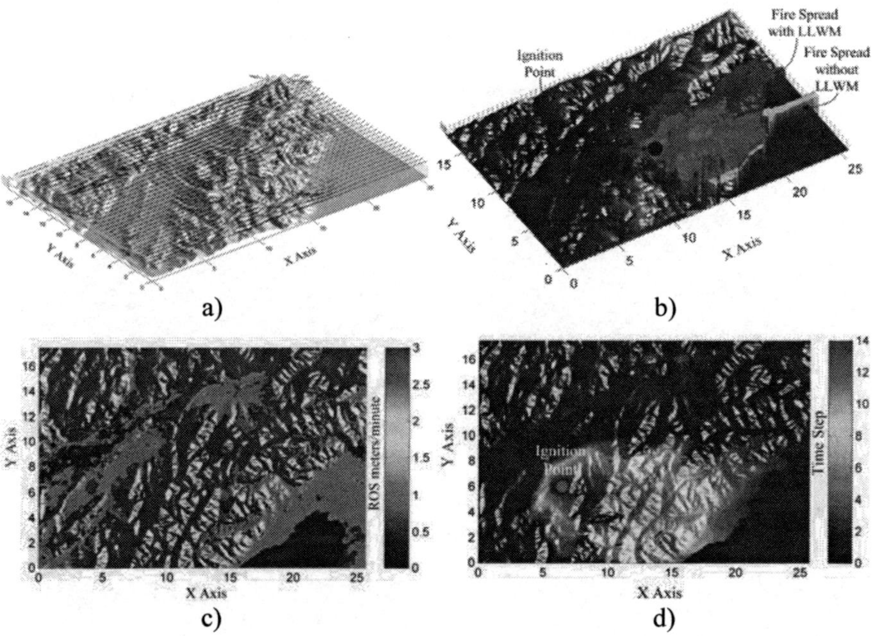

Figure 3: a) Wind vectors of the LLWM for west wind (coming from the left) of 20knt over the terrain, b) Comparison of fire spread, of CD-AUTH model, after time *t* assuming the LLWM and homogeneous wind c) The terrain and fuel characteristics used. d) CD-AUTH fire spread for different time steps coupled with LLWM.

In the above formulation *N* represents the eddy viscosity variable, C_b the surface friction coefficient, *u, v* the mean over the considered layer wind speed components in X and Y direction respectively, h the thickness of the atmospheric layer, defined by the minimum and maximum height of the terrain map and ζ the barometric pressure head distribution. The fire model is coupled with the wind 2DH boundary layer model, producing over the real topography the variable in intensity and direction wind field (its output is the wind speed and direction on every cell), enhancing the effects of ground relief. The model output comprises time sequences of the "cells' indices" matrix, allowing a subsequent estimation of the evolution of the fire front and the computation of rate of change of the burning and burnt areas during the fire event. The results of the LLWM are shown in Figure 3a) whereas in Figure 3b) the effect of taking into account the LLWM instead of homogenous wind to fire propagation is presented. The firespread for different time steps of the CD-AUTH model is presented in Figure 3 c) and d).

4 Comparison of the models

This section of the paper presents the comparison of the two models. For the purpose of our investigation 4 different time steps was chosen and these are represented by 10, 24, 36 and 48 hours after fire ignition. The simulation results are presented in Figure 4. The comparison represents the subtraction of the burnt

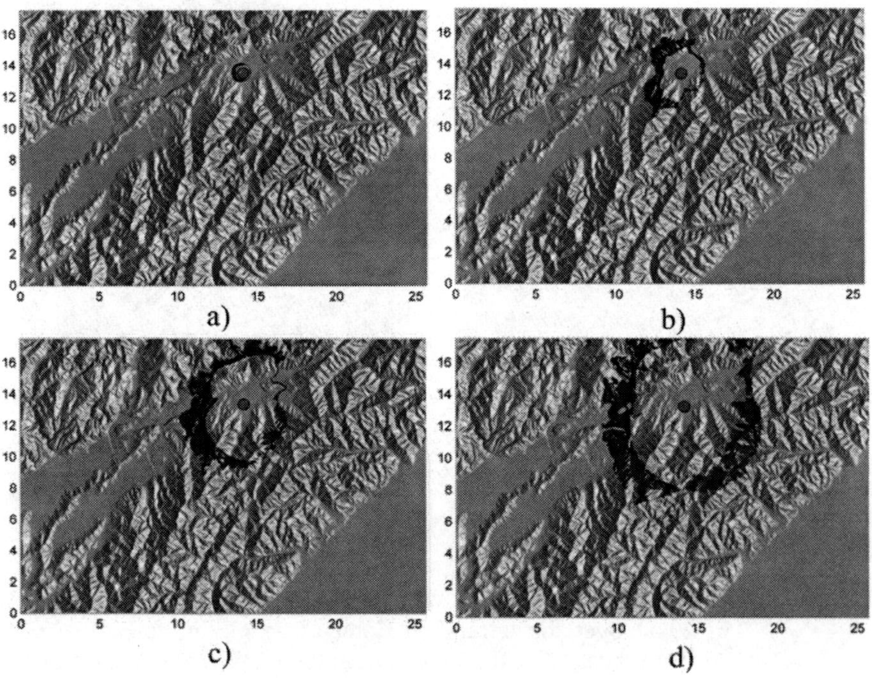

Figure 4: a) comparison for 10 hours after ignition, b) comparison for 24 hours after ignition, c) comparison for 36 hours after ignition, d) comparison for 48 hours after ignition. The red point represents the ignition point.

area computed by the Intesys model and the burnt area computed by the CD-AUTH model. As a result, the blue and red areas indicate the areas where the Intesys model predicted a faster fire line and delayed fire line respectively. It can be observed that the two models are in good agreement at the time step of 10 hours. This is because the fire spreads over a flat terrain and the predictions are mainly characterized by the fuel characteristics. On the other hand at greater time steps there is a small difference between the models and this is mainly caused to the modelling approximations and the shard terrain. In general, a good agreement is observed that is acceptable for fire predictions and management. The CPU demands of the two modelling approximation is identical but the CD-AUTH model presented higher demands when the low level wind module was introduced. This is because the solution of equation (7) in a numerical approximation requires high computation time.

5 WSN network topology control

Wireless sensor networks have been deployed for forest fire detection and monitoring. Furthermore, real time data sent by the sensor network is of vital importance to the fire management and improvement of the fire propagation models. WSNs usually provide weather characteristics that are related to FWI such as temperature, wind, humidity. These data are necessary for fire detection or monitoring of high risk locations. The time resolution that sensors send information to the manager varies according to weather characteristics. In case of low fire risk weather the sensors are set to idle mode in order to save energy whereas at high risk periods the sensors can send information every 15-30 minutes for early fire detection. This condition consumes considerable power and reduces the lifetime of the system. The sensors of the network are powered by a battery and energy efficiency is of vital importance. In [1, 15-16] energy saving techniques are presented based on routing and protocol implementations to wireless sensor networks. For the purpose of our investigation a WSN network topology control is developed that targets lifetime maximization. In case of a fire event, the CD-AUTH model is applied for fire spread predictions and this information feeds the sensor network to self manage and provide multi-time-resolution data of FWI to the fire manager. The proposed algorithm increases the sampling rate at the sensors that are placed on a zone of time T (Figure 5a)) around the current firefront without affecting the sampling rate of the rest sensors of the network. With this approach, fire managers are able to monitor in real time and with frequently updated data the fire event without wasting the total network energy. The goal is that sensors that are expected to be burned by the fire after time T are set to high, almost real time, sampling rates whereas the rest of the sensors monitor the area with the normal set values, providing energy efficiency. The algorithm implements the communication protocol presented in [16]. The power consumed for transmitting and receiving a message with r (bits/sec) over a distance d (m) is equal to

$$P_T(d) = (a_{11} + a_2 \cdot d^{\gamma}) \cdot r \quad P_R(d) = a_{12} \cdot r \tag{8}$$

Parameters a_{11}, a_2 are the transmitter electronic equipments (computational costs) and radio amplifier energy (communication costs) respectively. a_{12} represents the receiver electronic equipments and depends only on the computation processing. Parameter γ is the path loss exponent and depends on the communication link between each sensor and is usually set to 2 or 4. For the purpose of our investigation it was assumed that the sensors are separated by a space d=70m and γ=4. Parameter a_{11}= a_{12}= 50nJ/bit and a_2=0.0013pJ/bit/m^4 [16]. Two scenarios are compared. According to the first, in the case of a fire event the total network increases the bitrate from *Q bits/sec* to *W=3Q bits/sec* and is named as *P*. The second scenario concerns the implementation of the proposed topology control where the sensors placed in the area of interest (firefront after time T) increase their sampling rate and is denoted as P_{NC}.

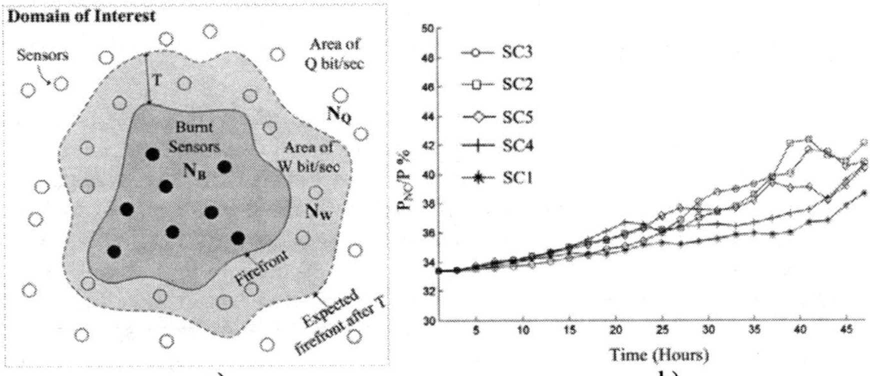

Figure 5: a) Interpretation of network topology algorithm. b) Power gain for 5 different fire scenarios over a period of 2 days after fire ignition.

An area of 5x5 Km was examined with total number of sensors N_T=*5100*. The total network power consumption for the two cases is computed according to

$$P = \sum_i^N \left(P_{Ti} + U_i \cdot P_{Ri}\right)_{\begin{cases} N=N_T-N_B \\ r=W \end{cases}}$$
$$P_{NC} = \sum_i^{N_Q} \left(P_{Ti} + U_i \cdot P_{Ri}\right)_{\begin{cases} N_Q=N_T-N_B-N_W \\ r=Q \end{cases}} + \sum_i^{N_W} \left(P_{Ti} + U_i \cdot P_{Ri}\right)_{\begin{cases} N_W=N_T-N_B-N_Q \\ r=W \end{cases}} \quad (9)$$

where N_Q is the number of sensors with bitrate *Q*, N_W is the number of sensors with bitrate *W*, N_B is the number of sensors burnt at time *t* and U_i is an on off parameters indicating if the sensor transmits only or if the sensor can receive and transmit data. The algorithm was implemented in 5 different scenarios and the simulation results are shown in Figure 5b). *SC1* represents the scenario where the ignition point was at the center of the terrain without wind. *SC2* and *SC3* represent the scenarios where the ignition point was at a west and east point of the terrain respectively with west and east wind velocities of 3m/s. Finally, *SC4* and *SC5* represent the scenarios where the ignition point was at north or south part of the terrain with winds blowing from north or south at 3m/s respectively. The high sampling rate time zone was assumed, corresponding to *T=2hours* and the fire was monitored for 2 days. It can be observed that P_{NC} is always less than P (by more than a fraction of 2) indicating the achieved energy efficiency of the proposed algorithm. It can also be observed that the power gain reduces with time. This is because the fire burnt area increases with time and so $N_Q \rightarrow N_W$. The power gain depends on the chosen high sampling rate zone (T) and the sampling rate (r).

6 Conclusions

This paper investigated two alternative fire modeling techniques based on CA (Intesys model) and cellDEVS (CD-AUTH model). It was shown that the CA method is characterized by less CPU demands and complexity but does not provide accurate results in windy conditions over sharp terrain. The CD-AUTH model was then used for network topology control of a WSN that target energy efficiency and high time resolution monitoring of forest fire. The effective operational use of the forest fire prediction model resulted to energy efficiency in the WSN of the order of 2.

Acknowledgement

This paper is part of the work driven by the EMMON (EMbedded MONitoring) ARTEMIS project.

References

[1] M. Hefeda, M. Bagheri, 'Forest fire modelling and early detection using wireless sensor networks', *Ad-Hoc & Sensor Wireless Networks*, vol. 7, pp. 169-224, Old City Publishing, 2009.

[2] W. J. de Groot, 'Interpreting the Canadian Forest Fire Weather Index (FWI) System', *in Proc. of Fourth Central Region Fire Weather Committee Scientific and Technical Seminar*, Canada, 1998.

[3] K. Pripuzic, H. Belani, M. Vukovic, 'Early forest fire detection with sensor networks: sliding windows sklylines approach', *Computer Science*, ISBN: 978-3-540-85562-0, Sringer, 2008.

[4] J. Lloret, M. Garcia, D. Bri, S. Sendra, 'A WSN deployment for rural and forest fire detection and verification', *Sensors*, 9, 8722-8747, 2009.

[5] L. Yu, N. Wang, X. Meng, 'Real time forest fire detection with WSN', *in Proc IEEE wireless communications, network and mobile computing*, vol. 2, 1214-1217, 2005.

[6] A. L. Sullivan, 'A review of wildland fire spread modelling, 1990-present 1: Physical and quasi-physical models', *Technical report, The Australian National University*, 2008.
[7] R. Rothermel, 'A mathematical model for predicting fire spread in wildland fuels', *Res. Pap. INT-115, U.S. Dept. of Agriculture-Forest service*, 1972.
[8] I. Karyfallidis, A. Thanailakis, 'A model for predicting forest fire spreading using cellular automoata', *Ecological Modeling*, 99: 87-97, 1997.
[9] P. Goncalves, P. Diogo, 'Forest fire modeling: A new methodology using cellular automata and geographic information systems', *in Proc. Int. Conf. on Forest Fire Research*, Nov. 1994.
[10] B. Malamud, D. Turcotte, 'Cellular automata models applied in natural hazards', *Computing in Science and Engineering*, 2:43-51, 2000.
[11] D. D'ambrosio, S. Di Gregorio, W. Spataro, G.A. Trunfio, 'A Model for the Simulation of Forest Fire Dynamics Using Cellular Automata', *in: Proc. of the iEMSs Third Biennial Meeting: "Summit on Environmental Modelling and Software"*, Burlington, USA, July 2006.
[12] L. Ntaimo, X. Hu, Y. Sun, 'DEVS-FIRE: Towards an integrated simulation environment for surface wildfire spread and containment', *Simulation*, vol. 84, no. 4, pp. 137-155, 2008.
[13] M. McLeod, R. Chreyh, G. Wainer, 'Improved Cell-DEVS models for fire spreading analysis', *Computer Science*, Springer, ISBN:978-3-540-40929-8, 2006.
[14] H. Anderson, 'Aids to determining fuel models for estimating fire behavior', *Tech. Rep. INT-122.USDA For. Serv.*, 1982.
[15] X. C. Nrahari, B. Simha, R. Cheng, M. X. Liu, 'Strong minimum energy topology in wireless sensor networks: NP-completeness and heuristics', *IEEE Trans. Mobile Comp.*, vol. 2, pp. 248-256., Sept. 2003.
[16] R. Mochaourab, W. Dargie,' A fair and energy efficient topology control protocol for wireless sensor networks', *Proc. Int. Conf. on Context-awareness for self managing systems*, pp. 6-15, 2008.

Diffusion limited propagation of burning fronts

M. Conti & U. M. B. Marconi
Dipartimento di Fisica, Universitá di Camerino, Italy

Abstract

In this numerical study we simulate burning propagation when the limiting transport mechanisms is the diffusion of oxygen and heat. This situation may be representative of smouldering combustion in the forest ground, when the intricate vegetation structure prevents the onset of large scale convection. The interplay between the diffusion processes results in a dynamical instability with a tendency of the burning front to develop cellular or dendritic patterns. The length scale associated with the observed patterns results to be a combination of the diffusion lengths associated with the two competing processes.
Keywords: flames propagation, pattern instability.

1 Introduction

Flame propagation is a complex process involving chemical reactions and transport phenomena [1]. The advancing of the combustion front is sustained by the self-produced heat and is rate limited by the availability of both fuel and oxidant. In general, the transport phenomena in the environment where the burning front propagates involve both heat and oxygen convection. However, in the smoldering combustion of the forest ground, when the intricate vegetation structure prevents the onset of large scale convection processes, diffusion may become the rate limiting mechanism. Diffusion limited growth is the situation observed in a variety of growth phenomena such as solidification processes, viscous fingering, electrochemical deposition, diffusion limited aggregation, dielectric breakdown [2]. In such phenomena the front of the growing phase is morphologically unstable and evolves into a complex pattern, with production of fingers, grooves, side-branches and dendritic structures. For many of these phenomena typical properties such as the velocity of the front and the length of the patterns can be related to well studied quantities such as surface tension, chemical potential differences,

WIT Transactions on Ecology and the Environment, Vol 137, © 2010 WIT Press
www.witpress.com, ISSN 1743-3541 (on-line)
doi:10.2495/FIVA100041

temperature gradients. This analogy stimulates interesting questions about the connection between the morphological instability of the burning front and the one related to dendritic growth in rapid solidification and similar phenomena.

As shown in the classical work of Mullins and Sekerka [3], the morphological instability observed in diffusion controlled growth arises as the growth process requires transportation of some conserved quantity away or towards the growing interface, and the fragmentation of the advancing front into a large surface area favors the diffusion processes. On the other side a large interface area is too costly in terms of surface energy, and the competition of these two effects determines the dynamics of the process and the characteristics of the interface pattern. Is this mechanism useful to give some insight into the combustion propagation problem? Is some factor playing the role of surface tension in this problem? We shall answer these questions using a simple lattice model [4] in which two diffusive fields, and a reactive field proposed on a purely phenomenological basis, interact and give rise to a rich variety of patterns.

The paper was stimulated by experimental studies performed by Zik et al. [5, 6] with a two-dimensional apparatus. These researchers used a thin rectangular chamber to study the combustion of a paper sheet. The thin gap between the two plates prevented convection transport of heat and oxygen. Measurements were made when the combustion was very slow-that is, the fuel was smoldering, a non flaming mode in which the emitted gas does not glow. In these conditions a fingering instability was observed.

The paper is organized as follows: in section 2 we introduce and motivate the model, in section 3 we present the results of the simulations of the lattice model for various choices of the control parameters, in section 4 we draw the conclusions.

2 A minimal model for combustion propagation

The propagation of a burning front is a complex phenomenon which involves several heterogeneous reactions. Details of the reactions kinetics are poorly understood, but some aspects of the global picture can be captured by a minimal model which refers to only a limited number of dynamic variables. In particular, the interfacial instabilities of the burning fronts observed in some experiments in two dimensions can be predicted and interpreted.

At a basic level of description, the combustion advances through the solid fuel in a competition between endothermic pyrolysis and exothermic oxidation. Then, the dynamics of the process is characterized only by three fields: the oxygen, the fuel and the heat released by the reaction.

The fuel reacts with oxygen and releases heat, in a local irreversible transformation from an unstable state before ignition towards a final stable state. On the other hand, oxygen and heat are transported by diffusion processes towards and away from the advancing front, respectively.

In the model we propose, the three fields are defined on a discrete two dimensional square lattice, where each location is labeled by i. Then, three dynamical variables A_i,C_i and H_i represent the oxygen concentration (A_i),the combustible C_i, which takes on the values 0 or 1 in the unburned or burned state respectively, and the heat amount H_i.

The dynamics of the three variables is developed along a discrete time grid, according to the following evolution rules:

1) a site i is chosen randomly .

2) if both the local heat and oxygen variables H_i and A_i are above some fixed thresholds (S_H and S_A) the discrete variable C_i will change its state ($0 \rightarrow 1$) in an irreversible fashion, representing the combustion of the site.

3) Due to combustion the site releases a certain amount of heat and consumes oxygen according to the following scheme:

$$C_i' = C_i + 1$$

$$H_i' = H_i + \Delta H$$

$$A_i' = A_i - \Delta A$$

4) The oxygen molecules diffuse towards the reaction zone, while the heat diffuses away from there. The two processes occur at different rates. The diffusion mechanism is included by assuming that at every time step random exchange events take place between the nearest neighbors sites i and j in the lattice, so that for each event the post-collision conserved field E (E stays for heat or oxygen) is determined according to the rule $E_i' \rightarrow (E_i + E_j)r$ and $E_j' \rightarrow (E_i + E_j)(1 - r)$ where r is a random number chosen from a uniform distribution in the unit interval. By sweeping randomly the lattice the system behaves ergodically and one can compute meaningful statistical averages.

This model is inspired to our earlier work [7], where solidification processes in binary alloys were described through a microscopic dynamics with stochastic character. In that case the process was reduced to a diffusion-reaction dynamics with two diffusive fields (temperature and solute concentration), recovering the observed macroscopic behavior at a microscopic and mesoscopic scale without coarse graining the model. However a main differences with respect to the solidification problem is that the combustion reaction is an irreversible process with no analogue in the solidification problem, where the material field may undergo either the liquid to solid transformation or the reverse.

Moreover, combustion has no analogue to the surface energy cost between neighboring sites belonging to different states, lacking a mechanism for the morphological stabilization of the burning front. The latter point raises the question whether some other factor plays the same role. To this regard we observe that the ignition at a site i is activated only if a sufficient amount of thermal energy (heat) is present. Heat is released by combustion of nearest burned sites,but is dispersed away by diffusion towards colder regions. However, if the associated thermal diffusion length is sufficiently short the result will be that of a stabilizing force similar to a surface tension, because only sites close to a site which is releasing heat can light up. On the other side the oxygen required for combustion at a given site must be transported there by diffusion. Thus the need for fresh oxygen tends to favor the formation of a large interface area. The balance between these two competing forces leads to a dynamical instability of the combustion front, whose characteristics will change according to the experimental conditions.

3 Numerical results

We performed simulation runs of the reactive-diffusive model described above in the xy plane, using a 512×512 lattice. In the initial configuration the sites are unburned ($C = 0$) and cold ($H = 0$), and the oxygen concentration is uniform ($A = A_0$). To start the combustion a thin strip ($x < x_0$) at the left border of the domain is prepared in the "hot" state ($H = 1$). In our simulations we fixed the oxygen concentration threshold at $S_A = 1$; the release of heat and the oxygen consumption are assumed to be $\Delta H = \Delta A = 1$. The combustion process was followed up to N Montecarlo steps (MCS), until a well defined regime was attained. The two diffusive time scales for the heat and oxygen fields were controlled by iterating independently, for each MCS, the related attempts of exchange. The resulting frequencies of attempt (per MCS) will be indicated in the following as f_H and f_O, respectively. To present the numerical results all lengths will be measured in lattice spacing units and the non-dimensional time will be expressed in MCS.

The structures developed in the combustion process strongly depend on the diffusion rate of the oxygen field. In Figure 1 we show, at $t = 16000$, the cellular pattern which arises fixing $f_H = 0.2$, the oxygen concentration at $A_O = 0.90$ and

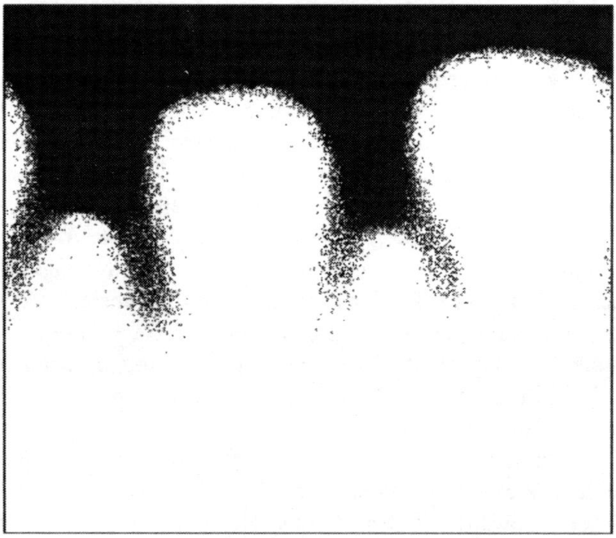

Figure 1: The combustion field at $t = 16000$ MCS. The initial oxygen concentration is $A_0 = 0.90$, the frequencies of attempt for the heat and the oxygen fields are $f_H = 0.2$ and $f_O = 1$ respectively, the heat threshold is $S_H = 0.50$.

the heat threshold at $S_H = 0.50$. The frequency of attempt f_O (that means the diffusivity of the oxygen field) is set at $f_O = 1$. Similar structures are observed in the directional solidification of binary alloys beyond the onset of the Mullins-Sekerka instability. The characteristic length of the pattern is selected through the competition of the stabilizing effect of the surface tension and the necessity to develop a large interface area to reject (or to draw) a conserved quantity.

Both these phenomena can be described through natural length scales (the capillary length d_0 and the diffusion length l_d, respectively), and the wavelength of the pattern emerges as $\lambda \sim \sqrt{d_0 l_d}$. Our model does not account for an interface energy cost, nevertheless a stabilizing effect, mimicking the role of surface tension, is still present, as the ignition is prevented when the local heat field is below the threshold S_H. But the heat released at a burning site is dissipated through a diffusion mechanism, in such a way that too fragmented structures are disfavoured.

According to the above considerations, we expect a thinner morphology of the combustion pattern with decreasing either the threshold S_H or the oxygen diffusion length L_O (that means decreasing the frequency of attempt f_O). This is the situation depicted in Figures 2 and 3. In Figure 2 ($f_O = 0.2$)we observe that the wavelength of the pattern has been strongly decreased, as we here observe five well developed cells of the burned fuel. In Figure 3 ($f_O = 0.2$, $S_H = 0.40$)the situation is even more clear, as we observe tip splitting and a side-branch activity that indicates that the pattern is driven towards a dendritic regime.

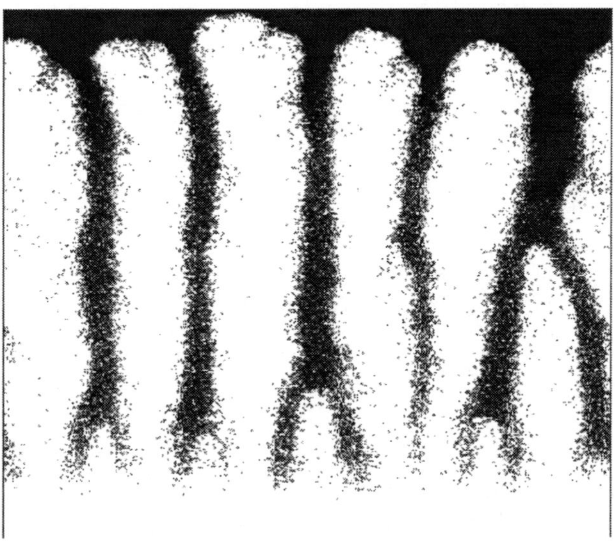

Figure 2: The combustion field at $t = 16000$ MCS. The initial oxygen concentration is $A_0 = 0.90$,the frequencies of attempt for the heat and the oxygen fields are $f_H = 0.2$ and $f_O = 0.2$ respectively, the heat threshold is $S_H = 0\ 50$

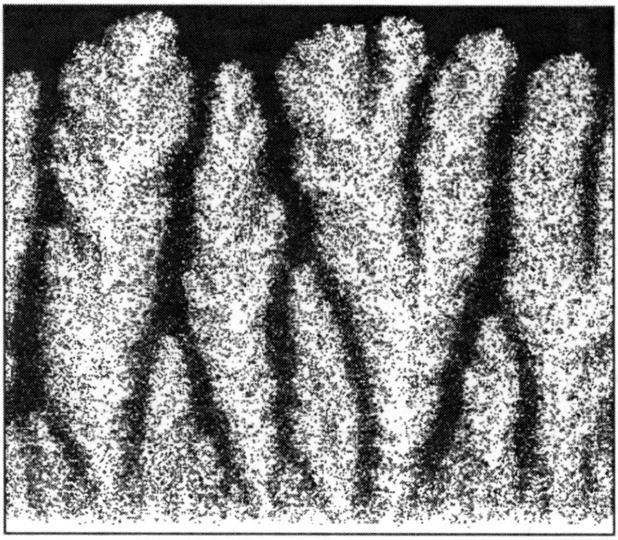

Figure 3: The combustion field at $t = 16000$ MCS. The initial oxygen concentration is $A_0 = 0.90$,the frequencies of attempt for the heat and the oxygen fields are $f_H = 0.2$ and $f_O = 0.2$ respectively, the heat threshold is $S_H = 0.40$.

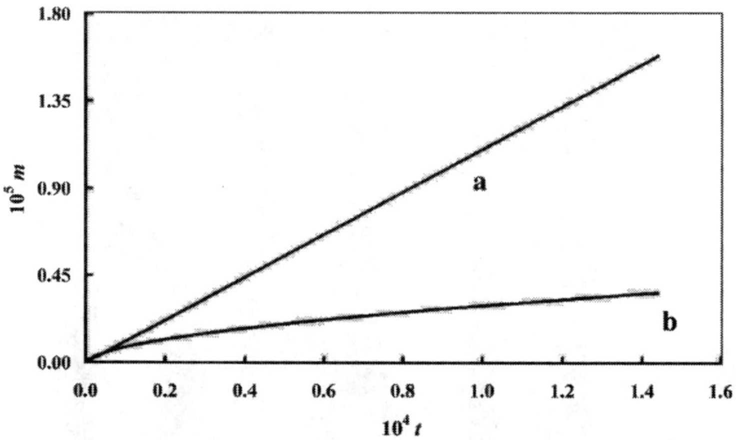

Figure 4: The mass of the combusted sites represented versus time. Curves a and b refer to cellular and compact growth, respectively.

Notice that the concentration of oxygen is initially fixed at a value below the threshold S_A, and to sustain the combustion oxygen must be drawn towards the interface. Then, a compact front would be slowed down with the growth rate

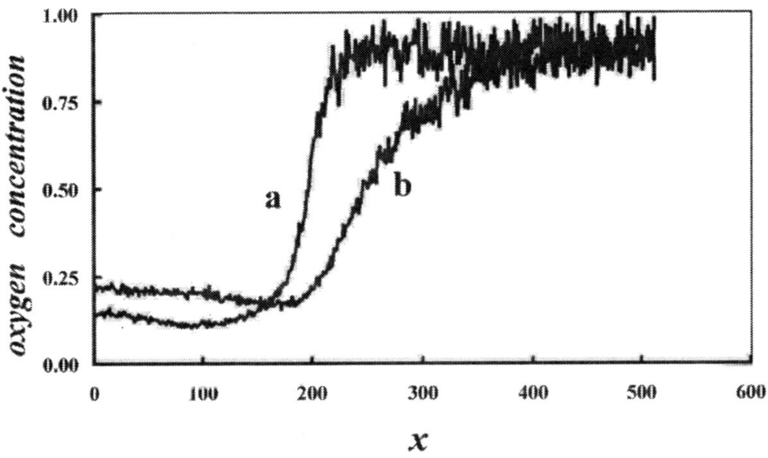

Figure 5: The profile of the oxygen field for the cellular growth shown in fig.1 (curve b) and in fig.2 (curve a).

decaying as $t^{-1/2}$. But a cellular or dendritic structure can develop at constant growth rate if the burnt sites left behind the advancing front cover an area fraction equal to the initial oxygen concentration. This is precisely the situation shown in Figure 4, where the "mass" of the combusted sites is represented versus time. The straight line (curve a) refers to the cellular growth addressed in Figure 2: we can observe that the combustion front advances at constant growth rate. For comparison we also show a curve (b) representative of compact growth conditions obtained with $f_O = 1, S_H = 0.45, A_0 = 0.60$. In the latter case we observe a typical diffusion-limited behavior, with the combustion front advancing as $\sim t^{\alpha}$: the deviation of the best fit value $\alpha = 0.53$ from the pure diffusional value $\alpha = 0.5$ can be attributed to a residual porosity of the combusted phase.

We observed that the morphology of the combusted phase is strongly affected by the length scale of the oxygen field. The latter in turn depends on the frequency of attempts f_O, which is the microscopic counterpart of the macroscopic diffusion coefficient D. To elucidate this point we show in Figure 5 the oxygen concentration along the growth direction, at $t = 8000$. The two curves refer to the same sets of data employed in Figs. 1 and 2, and the oxygen field is averaged over the direction normal to the combustion propagation. We note that the width of the transition zone from the low concentration area (behind the combustion front) to the high concentration sites increases with increasing the oxygen diffusivity f_O.

Figure 6 shows the diffusion length L_O of the oxygen field versus f_O. The two curves refer (from top to down) to $S_H = 0.5$ and $S_H = 0.3$. L_O has been estimated as the length required for the transition from 20% to 90% of the concentration at infinity. Notice that a larger value of S_H reflects on a lower velocity of the process, and we recover the well known result that the diffusion length diminishes as the growth rate increases.

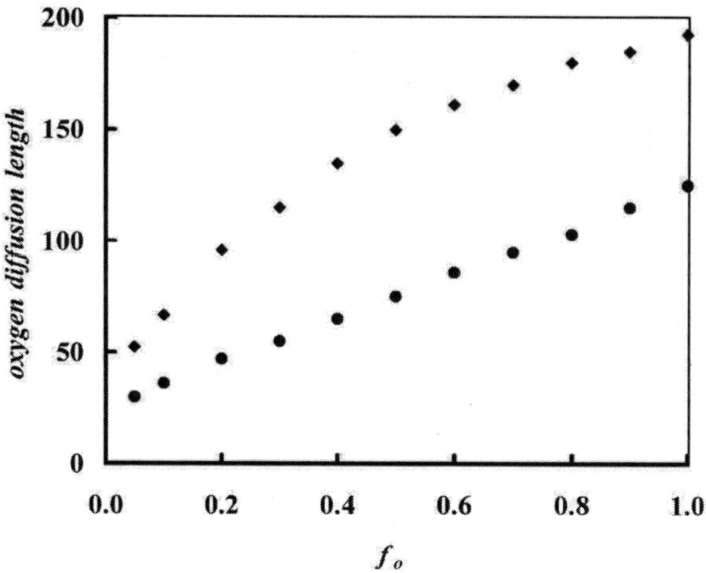

Figure 6: The diffusion length of the oxygen field versus the frequency of attempts f_O. The two curves were obtained with $S_H = 0.5$ (diamonds) and $S_H = 0.3$ (circles).

The above considerations suggest that the propagation of a burning front could be described,in a S_H, f_O plane, through a morphological phase diagram, where thinner structures (corresponding to dendritic-like patterns of the combustion front) should correspond to a large growth rate or low oxygen diffusivity, whereas cellular patterns are likely to be found at large values of S_H and f_O. This kind of diagram is shown in Figure 7, for an initial oxygen concentration $A_0 = 0.75$.

4 Conclusions

In the present paper we presented some numerical results obtained with a lattice model which describes the propagation of combustion in the absence of convection. We observed that the diffusive transport of oxygen is at the origin of a morphological instability of the combustion front. The process lacks a stabilization mechanism at the microscopic level, however the necessity to preserve adequate temperature conditions for the burning reaction results in a tendency to minimize the area of the advancing front, mimicking the effects of surface tension. The resulting pattern is characterized by a typical length scale which is related to the diffusion length of the oxygen and the heat fields. Even in partial defect of oxygen, the burning front can advance at constant growth rate as unburned fuel is left behind the advancing front. The numerical simulations carried out at various

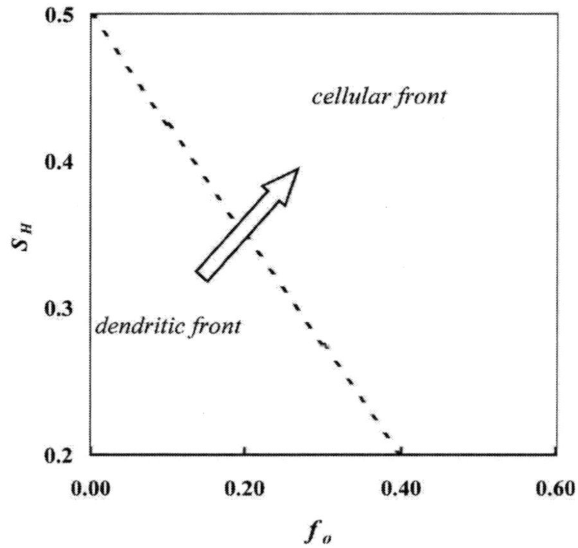

Figure 7: The morphological phase diagram for the pattern of the combustion field in the S_H, f_O plane.

diffusion rates display the complex morphology of the interfacial patterns and allowed us to construct a phase diagram of the phenomenon.

References

[1] I. Glassman and R.A. Yetter, *Combustion* (Academic Press, Burlington MA, 2008).

[2] J.S. Langer, *Instabilities and pattern formation in crystal growth*, Rev. Mod. Phys. 52,(1980), pp. 1–28.

[3] W.W. Mullins, R.F. Sekerka *Morphological stability of a particle growing by diffusion and heat flow*, J. Appl. Phys, 34 (1963),pp. 323–329.

[4] M. Conti, U. Marini Bettolo Marconi *Fingering in slow combustion*, Physica A, 312 (2002), pp. 381–391.

[5] O. Zik, Z. Olami, E. Moses, *Fingering instability in combustion*, Phys. Rev. Lett. 81 (1998), pp. 3868–3871.

[6] O. Zik, E. Moses, *Fingering instability in combustion:an extended view*, Phys. Rev. E 60 (1999), pp. 518–531.

[7] M. Conti, U. Marini Bettolo Marconi *Novel Monte-Carlo lattice approach to rapid directional solidification of binary alloys*, Physica A, 277 (2000), pp. 35–46.

www.witpress.com, ISSN 1743-3541 (on-line)

Statistical parameter estimation for a cellular automata wildfire model based on satellite observations

E. Couce & W. Knorr
Department of Earth Sciences, University of Bristol, UK

Abstract

The importance of understanding the impact of wildfires on natural ecosystems has given rise to the development of realistic computer models for the simulation of wildfires. Stochastic models based on simplified equations and local interactions, such as Cellular Automata (CA) models, are particularly popular as an alternative to more computationally demanding deterministic models. However, the challenges associated with observing wildfires under natural conditions, and the highly non-linear nature of fire spread makes it extremely difficult to parameterize them. In this work we present a method for adjusting the behaviour of one such CA model from the statistical analysis of satellite data of more than 750,000 African wildfires detected in 2003. Statistical metrics are developed to characterize agreement between model and satellite observations. The average probability of fire transmission amongst cells and the spatial scale of the model are adjusted so that maximum agreement is found between model output and the observed extension and statistical distribution of the real fires. While the results obtained are only valid for the particular CA model used and within the geographical limits of the region studied, we believe the process could be adapted to fine-tune and validate other CA models in regions where enough fire observations are available.

Keywords: fire spread model, cellular automata, parameter estimation, African savanna wildfires, satellite observations.

1 Introduction

The importance of wildfires for natural ecosystems, together with the socio-economic danger they represent, have lead to a great deal of effort invested in the

www.witpress.com, ISSN 1743-3541 (on-line)
doi:10.2495/FIVA100051

simulation and modelling of fire behaviour. There is strong demand for accurate fire simulations that would provide an important tool for fire fighters and other people involved in fire management. However the modelling and understanding of wildfires is a highly complex problem, governed by non-linear equations and depending on more factors than can possibly be known at any time and which affect the fire behaviour in non-trivial ways. Because of this, deterministic models based on theoretical partial differential equations have only found limited success in the description of wildfires, and this only at the cost of large amounts of computer resources and processing time [1, 2]. Thus, stochastic models based on simplified empirical or semi-empirical equations have remained a popular alternative.

One of the main obstacles to the development of a realistic model for the simulation of the spread of wildfires lays in its validation with data from real fires. This is mainly due to the fact that the knowledge of the factors required by the models (such as humidity, wind information and vegetation type and state) is limited for the wildfires on record. Many published fire spread models are not compared directly to real data. Instead, a model's validity is often discussed by comparing its output to that of some better-known model, usually for only one or two particular cases. If real data are used in the comparison, they are typically from a single wildfire and of high spatial and temporal resolution. While this is no doubt a very relevant initial test, it can hardly be considered enough for a process as complex as wildfire spread, particularly if the model is to be applied to a wider range of conditions.

In this work we compare the output of a new stochastic CA model for the spread of wildfires with statistics of the areas of real fires detected by satellite within an extensive region of Africa during the 2003 fire season (nearly 800,000 wildfires). The model's parameters are initialized with observations of vegetation type, wind, temperature, precipitation, and FAPAR (fraction of plant-absorbed photosynthetically active radiation) from the study region. They are then adjusted to better reproduce the histogram of the observed fire areas, in order to establish the model's optimal spatial resolution and average probability for fire spread. While this technique does not replace the need to compare a model's output with data of high spatial and temporal resolution, we believe it represents a significant step towards comprehensive validation of fire spread models.

2 Methods

2.1 Fire data

The fire data used in our study was obtained by the Moderate Resolution Imaging Spectroradiometer (MODIS) carried by the Terra (EOS AM) and Aqua (EOS PM) satellites, with a spatial resolution of 500x500 m^2. The study region (Figure 1) corresponds to the MODIS tiles h19v10 and h20v10, an area of approximately 1200x2400 km^2, with latitude spanning 10.00ºS to 20.00ºS, and longitude between 10.15°E and 31.92°E. It includes portions of Angola, Zambia, Namibia, Botswana, Zimbabwe, Congo, Zaire and Mozambique. Data were available for the fire seasons (April to November) of 2000 to 2004.

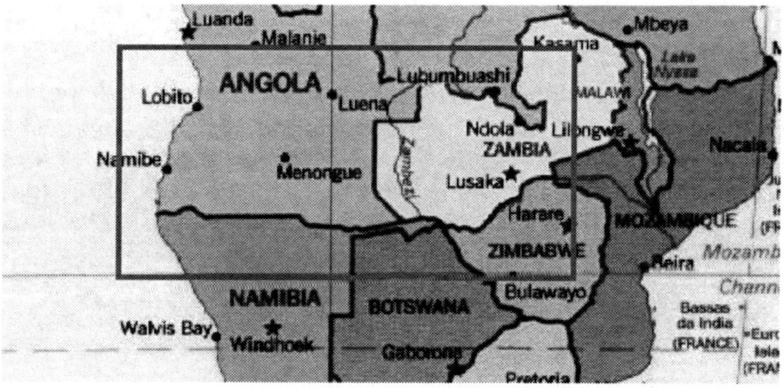

Figure 1: Study region, indicated by rectangle.

Fires were identified with a generic algorithm being developed by Rebelo et al. [3] and Rebelo [4]. This algorithm detects areas exhibiting sudden changes based on discrepancies between expected and observed bi-direction reflectance (BRDF) observations. In the study region, one of the most common causes of these sudden changes are wildfires, although several additional tests are used to separate burning from phenological changes within the pixel. The algorithm is similar to that of the MODIS fire product (see Roy *et al.* [5, 6]).

The algorithm detects the day a sudden change in BRDR suggests the onset of a wildfire in the region. Therefore the only information available is the probable day a fire starts for a pixel, and nothing is known about the fire's duration. Nevertheless considering the type of vegetation present in the study region, predominantly savanna and shrubland, these wildfires would tend to propagate fast and have short duration.

2.2 Additional geographical information

The Type 2 MODIS Land Cover Product (MOD12Q1) was used to identify types of vegetation present within the study region [7, 8]. Five different types of vegetation were considered in the study: savanna, woody savanna, grassland, open shrubland and deciduous broadleaf forest. Together they encompass over 80% of the total area analyzed, with savanna and woody savanna representing 45% and 20% respectively.

In order to estimate the potential amount of burnable fuel available in each grid cell, we employed the monthly, gridded 0.5° by 0.5° FAPAR product of Gobron *et al.* [9] for 2002-2003, generated from an analysis of the data recorded by the Sea-viewing Wide Field-of-view Sensor (SeaWiFS).

Some wind information was available for the period from February 1999 to October 2001 from measurements carried out by a station of the FLUXNET network located near Maun, Botswana [10, 11].

For the climate input we used daily precipitation and daily minimum and maximum temperatures. The values of those variables and of solar incoming radiation for the period 2000 to 2004 were generated on a global 2° latitude by 2° longitude grid using the method of Nijssen *et al.* [12], based on daily station data from the Summary of the Day Observations (Global CEAS), National Climatic Data Center, and monthly gridded data. Monthly gridded temperature was obtained from the data set of Jones *et al.* [13, 14], with gaps filled from data of Hansen *et al.* [15, 16]. Monthly gridded precipitation data came from a 1.0° version of Chen *et al.* [17].

2.3 The model

The model used in the study [18] is laid out on a rectangular 2-dimensional lattice. It takes the cells initially on fire as input, and reproduces the possible evolution of the fire over successive time steps. The fire spread relies on the computation of semi-empirical probabilities of fire transmission from cells on fire and is stochastic in nature. The probabilities are computed based on climate factors, vegetation type, wind intensity, topography, fire intensities, and fuel content of each cell. Although the main form of propagation occurs among neighbouring cells, propagation from other cells is also possible by the emission of sparks, influenced both by wind and topographic conditions. The duration of the fire on each cell is computed from the amount of fuel existing on the cell and the fire's intensity. The model does not present spurious symmetry, and the results obtained appear realistic and successfully reproduce features of real wildfires, such as spotting.

For this analysis, the probability of transmission was expressed as the product of independent factors reflecting the effects of vegetation type, climate, wind strength, and average fuel load respectively. The effect of the vegetation type was implemented by re-scaling the percentages of the area burned within each vegetation type against data from a previous exhaustive study of the region during 2000 to 2004 (Roy *et al.* [5]). Climatic factors affecting fire spread make use of the widely adapted Nesterov Index [19], which takes into account the maximum daily temperatures of any series of consecutive days without significant rainfall. A preliminary analysis of the fire behavior in the area for the period for which wind data were available suggested a relationship between the probability of transmission and the maximum daily wind strength. Finally, the average litter load in the region of interest was approximated as the integral of the losses in leaf mass, estimated from time-integrated decreases in satellite-derived FAPAR.

2.4 Methods

Since the spatial resolution of the fire data (250,000 m^2 per pixel) is much coarser than the model's lattice cells, the comparison with the model output was carried out statistically considering the total burned area of each pixel. Burning pixels were considered to belong to the same fire if they were both contiguous in space (neighbouring cells) and time (the changes must have been detected either the same day, or with one day difference). A histogram with the fire areas of the nearly 800,000 fires detected in 2003 was compared with histograms obtained from a large number of model simulations.

For each model run, a fire starting point within the study area and a start data between April to November of 2003, the period for which all required data were available, were chosen at random. The random election of a date is justified, since the amount of detected fires barely changed between the different months of the fire season analyzed (not so the number of burning pixels, which clearly peaks in July and August, implying more extensive fires during those months).

The initial distribution of fuel load on the model lattice, i.e. at a much higher spatial resolution than the available satellite observations, was again generated through a random process for each run. The average fuel values were thereby kept below 30%, giving rise to fast, low intensity wildfires, similar to the ones observed for typical savanna conditions.

The contributions of type of vegetation, climate, and fuel load to the average probability of transmission were computed as a function of vegetation type, litter load, and Nesterov index. For the computation of the effect of the wind, a random value for the maximum daily wind strength was generated, based on the monthly averages and standard deviations values measured at the Botswana station.

The final probability of fire transmission between cells was obtained from the product of the contributions of vegetation type, climate factor, and fuel load multiplied by a global constant k, which was optimised by maximizing agreement between modelled and observed fire extension histograms. k was allowed to take on several values within the interval $(0, 1)$. The model was run multiple times for each value of k. Over 350,000 simulations were run in total, all employing a 100x100 lattice.

The equivalence between the number of cells that are counted as burned in a simulation and a burned pixel from the satellite fire data depends both on the spatial resolution of the model (which was allowed to change) and the percentage of the area of a pixel that needs to burn in order for the fire to be detected. For the satellite fire detection algorithm employed, the latter falls somewhere between 10 and 20%, and in this study it was set at 15% for comparison with the model output.

3 Results

The comparison between the histogram derived from observations and model output was carried out using the Kullback-Leibler divergence [20], which has previously been applied to the testing of other ecological models [21, 22].

The number of bins for the histogram is limited by the number of satellite data pixels that correspond to the area represented by the model's 100x100 lattice. We found an optimal representation at seven bins, considering number of events per bin and number of data points of the two histograms.

Figure 2 shows the variation of the Kullback-Leibler divergence as a function of the model's resolution (A, B) and the value of *k* (A, C). We can see that the KL divergence value does not vary much with resolution as long as it stays above ca. 22–23 meters, although the fit does get progressively worse as the spatial resolution gets coarser. The dependence on *k* is much stronger, with a very well defined minimum at 0.46.

The optimal model histogram (corresponding to *k*=0.46 and a spatial resolution of 26x26 m^2) is shown in Figure 3 for comparison with that of the satellite observations from the study area (in black). We find good agreement between the histograms, with a final value for the Kullback-Leibler divergence of 0.015.

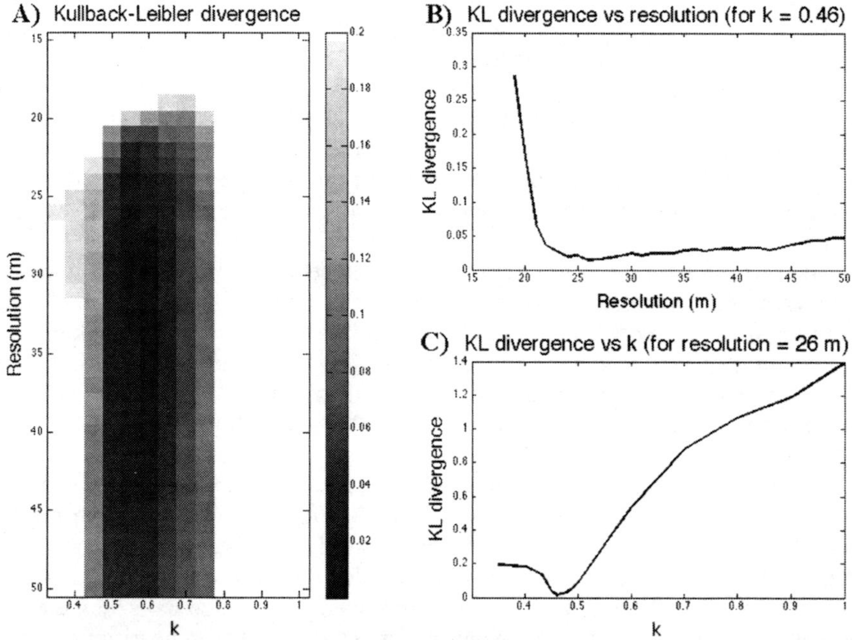

Figure 2: A) Kullback-Leibler divergence as a function of the probability factor *k* and the model's spatial resolution (indicated by the length corresponding to the side of a cell). The minimum is found for *k*=0.46 and a resolution of 26x26 m^2. B) KL divergence as a function of the spatial resolution, for *k*=0.46. C) KL divergence as a function of *k*, for a resolution of 26x26 m^2.

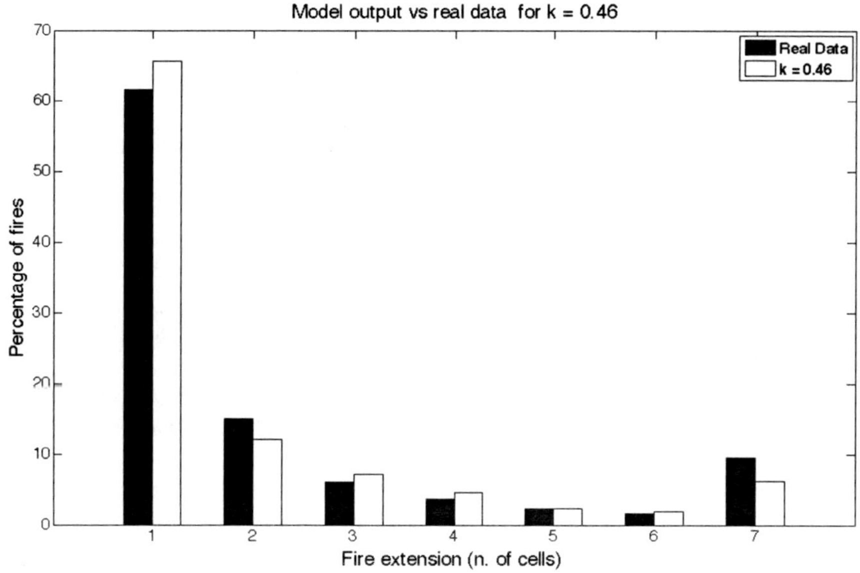

Figure 3: Comparison of the histograms with the percentage of fires detected as a function of their extension, for real data (black) and the optimal model output (white), for k=0.46 and a spatial resolution of 26x26 m^2. Fire extension is measured in pixels, each corresponding to 250,000 m^2. For these two distributions the Kullback-Leibler divergence reaches the minimum value of 0.015.

4 Conclusions

In this work we have presented a new stochastic model for the prediction of the spread of wildfires based on cellular automata on a square grid, and its application to African savanna fires. After running multiple simulations with random conditions from the extensive study region, the fire area distribution of the model was compared to the satellite-derived fire data for the fire season of 2003 for a large region of southern Africa, with over 750,000 detected wildfires. The method has allowed the selection of the optimal average probability of fire spread and spatial resolution of the model. A good agreement has been found, with a value for the Kullback-Leibler divergence of 0.015 for a 7-bin histogram of the frequency each fire area was detected/simulated.

We believe this result lend valuable credibility to the model in an extensive set of conditions, in particular given that fire spread models are rarely compared to such an extensive set of real wildfire data. However it does not eliminate the need of further tests. The direct comparison of model output with the evolution of real fires with data available at high spatial and temporal resolution would greatly benefit the adjustment of the model's behaviour, particularly regarding the effects of topography and wind on the probability of transmission.

Acknowledgements

We wish to thank Philip Lewis and José Gómez Dans for providing the fire data, and Almut Arneth for the wind data. This work greatly benefitted from discussions with José Gómez Dans, George Pepotroulos, and Jordi Burguet Castell.

References

[1] Bradley, J.H. & Clymer, A.B., Difficulties in the simulation of wildfires. *1993 International Emergency Management and Engineering Conference*: pp. 161–171, 1984.

[2] Karafyllidis, I. & Thanailakis, A., A model for predicting forest fire spreading using cellular automata. *Ecological Modelling*, **99(1)**, pp. 87–97, 1997.

[3] Rebelo, L., Lewis, P., et al., A temporal-BRDF model-based approach to change detection. *Geoscience and Remote Sensing Symposium, IGARSS '04*, 2004.

[4] Rebelo, L.M., The development of a generic change detection scheme: An application to the identification and delineation of fire affected areas. PhD thesis, Department of Geography, Remote Sensing Unit, London, University College London, 2005.

[5] Roy, D.P., Lewis, P.E., et al., Burned area mapping using multi-temporal moderate spatial resolution data–a bi-directional reflectance model-based expectation approach. *Remote Sensing of Environment*, **83(1–2)**, pp. 263–286, 2002.

[6] Roy, D.P., Jin, Y., et al., Prototyping a global algorithm for systematic fire-affected area mapping using MODIS time series data. *Remote Sensing of Environment*, **97(2)**, pp. 137–162, 2005.

[7] Strahler, A., Muchoney, D., et al., MODIS Land cover product: Algorithm Theoretical Basis Document. 1999.

[8] Hansen, M.C., Defries, R.S., et al., Global land cover classification at 1km spatial resolution using a classification tree approach. *International Journal of Remote Sensing*, **21(6–7)**, pp. 1331–1364, 2000.

[9] Gobron, N., Pinty, B., et al., Evaluation of fraction of absorbed photosynthetically active radiation products for different canopy radiation transfer regimes: Methodology and results using Joint Research Center products derived from SeaWiFS against ground-based estimations. *Journal of Geophysical Research-Atmospheres*, **111(D13)**, 2006.

[10] Veenendaal, E.M., Kolle, O., et al., Seasonal variation in energy fluxes and carbon dioxide exchange for a broad-leaved semi-arid savanna (Mopane woodland) in Southern Africa. *Global Change Biology*, **10(3)**, pp. 318–328, 2004.

[11] Arneth, A., Veenendaal, E.M., et al., Water use strategies and ecosystem-atmosphere exchange of CO2 in two highly seasonal environments. *Biogeosciences*, **3(4)**, pp. 421–437, 2006.

[12] Nijssen, B., Schnur, R., et al., Global retrospective estimation of soil moisture using the VIC land surface model, 1980–1993. *Journal of Climate*, **14**, pp. 1790–1808, 2001.
[13] Jones, P.D., New, M., et al., Surface air temperature and its variations over the last 150 years. *Reviews of Geophysics*, **37**, pp. 173–199, 1999.
[14] Jones, P.D., Osborn, T.J., et al., Adjusting for sampling density in grid box land and ocean surface temperature time series. *Journal of Geophysical Research-Atmospheres*, **106(D4)**, pp. 3371–3380, 2001.
[15] Hansen, J., Ruedy, R., et al., GISS analysis of surface temperature change. *Journal of Geophysical Research-Atmospheres* **104(D24)**, pp. 30997–31022, 1999.
[16] Hansen, J., Ruedy, R., et al., A closer look at United States and global surface temperature change. *Journal of Geophysical Research-Atmospheres*, **106(D20)**, pp. 23947–23963, 2001.
[17] Chen, M.Y., Xie, P.P., et al., Global land precipitation: A 50-yr monthly analysis based on gauge observations. *Journal of Hydrometeorology*, **3(3)**, pp. 249–266, 2002.
[18] Couce, E., A stochastic cellular automata model for the spread of wildfires: casestudy of the African savanna fires. MSc thesis, Dept. of Earth Sciences, University of Bristol, Bristol, 2008.
[19] Nesterov, V.G., Fire Frequency Index and Methods of Its Determination, Goslesbumaga, Moscow, 1949.
[20] Kullback, S. & Leibler, R.A., On information and sufficiency. *Annals of Mathematical Statistics*, **22(1)**, pp. 79–86, 1951.
[21] Burnham, K.P. & Anderson, D.R., Kullback-Leibler information as a basis for strong inference in ecological studies. *Wildlife Research,* **28(2)**, pp. 111–119, 2001.
[22] Burnham, K.P. & Anderson, D.R., *Model Selection and Multimodel Inference: A Practical Information-Theoretic Approach, Second Edition,* Springer Science, New York, 2002.

Sand on fire: an interactive tangible 3D platform for the modeling and management of wildfires

S. Guerin[1,2] & F. Carrera[3]
[1]*Simtable LLC, USA*
[2]*Santa Fe Complex (sf_X), USA*
[3]*Worcester Polytechnic Institute (WPI), USA*

Abstract

This paper presents the current development of an interactive tangible 3D platform that is used to conduct wildfire training, incident command and community outreach activities by allowing one to interactively visualize a variety of scenarios on sand tables, based on underlying wildfire, traffic, smoke, rain and incident command models. The platform, which is part of a larger effort to create ambient interactive environments at the Santa Fe Complex, consists of a coordinated camera-projector pair that uses active surface projections to detect physical interaction on an actual sand table. Our hardware and software create simulations on sand surfaces, where the changeable 3D surface is an active part of the simulation. By decoupling the sensing of physical interactivity from the underlying models, our platform is model-agnostic and could be used to visualize fire propagation and evacuation models from a variety of sources. Its value lies primarily in the immediate reactivity of the touchable sand surface, which engages users more intimately than other traditional training and education tools.

Keywords: sand table, interactive, tangible, platform, wildfire modeling, wildfire evacuation, emergency planning, traffic simulation, advanced visualizations, firefighter training.

WIT Transactions on Ecology and the Environment, Vol 137, © 2010 WIT Press
www.witpress.com, ISSN 1743-3541 (on-line)
doi:10.2495/FIVA100061

1 Background

The City of Santa Fe, New Mexico (USA) faces a wildland fire threat, as do many other communities in the world. In the City there are significant numbers of homes nestled in wild vegetation along narrow roads. Often there is only one ingress/egress to those neighborhoods. Fire Department officials began wondering in 2003 about whether residents would be able to evacuate their homes safely in the event of a fast-moving wildfire, and they began teaching citizens to be ready to evacuate or shelter-in-place depending on the circumstances.

In 2004, Santa Fe City officials initiated a collaboration with agent-based modelers and visualization scientists currently at the Santa Fe Complex and with fire scientists at Anchor Point Group of Boulder, CO to develop models of the interaction between massive but not unrealistic wildfires and traffic as it would exist in an evacuation (the 48,000 acre Cerro Grande Fire, which destroyed over 200 homes in Los Alamos, was visible from Santa Fe). The goals were not only disaster-response planning but also education of citizens and first responders, as shown in fig. 1. City emergency response professionals believe that the ability to visualize a fast-moving fire and its accompanying smoke and their dramatic effect on traffic will serve as an effective educational tool and as a means to illuminate decision-making [1].

Since its beginnings [1], the interactive platform has evolved considerably and has found a variety of applications, including the simulation of boat traffic in the canals of Venice, Italy [2].

More recently (2009) a Santa Fe Complex spin-off company has begun packaging a sand table product (fig. 2) including a series of training tools aimed primarily at firefighting academies [3].

Figure 1: Traditional "table top" exercises used by multi-agency emergency planners.

Figure 2: The Simtable system [3].

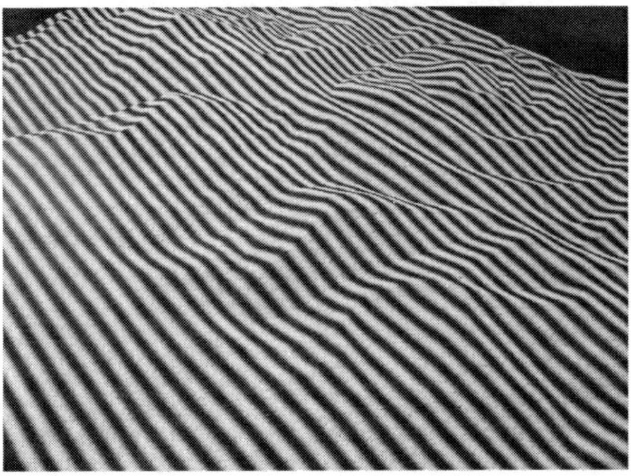

Figure 3: Structured light projected on the sand surface read by a webcam for elevation measurement.

2 The interactive tangible 3D platform

The platform consists of a physical sandbox, a computer projector, and a web camera coordinated by an operating system software. The webcam is used to sense human computer interactions and, in combination with the projector, to perform two "*structured light*" scans in order to establish:

1. a correspondence from camera pixels to projector pixels and
2. the height of the sand in real time (fig. 3).

The elevation scan allows the platform to provide visual feedback to users to guide them as they reconstruct a specific landscape based on topographic maps.

The platform's operating system detects interactions within the camera's field of vision and uses Open Sound Control (OSC) protocols to communicate with separate modeling applications running on the platform. Another spinoff of the Santa Fe Complex is developing an application programming interface (API) that will allow any programmer to write applications for the many other potential uses of the platform [4].

Current applications of the platform [3] are designed to react to laserpointers, whereas laboratory versions can also react to hand or facial gestures and can incorporate physical fiducials to represent various types of interactions, as was done with the Venice Traffic Table [2].

The 3D elevation detection algorithms based on the *structured light* approach, our platform allows a user to select a geographical area in Google Earth and follow color cues to form the elevation map in the sand.

Digital Elevation Models (DEMs) are loaded from a seamless dataset which provides 10 meter resolution DEMs for the US and 30 meter resolution DEMs for locations outside of the US. Data can also be loaded from geoservers supporting the Web Mapping Service (WMS) and Web Feature Service (WFS) protocols. DEM files can be loaded from local government sources if higher resolutions are needed. It is often the case that local governments have higher resolution files resulting from LIDAR surveys or other sources.

Once a DEM is loaded, the Simtable scans the current height of the sand for comparison and colors the sand red where sand needs to be removed and colors green where sand needs to be placed. This "sculpt by color" allows users to quickly form an area of interest in less than 20 seconds.

Further realism is added to the sand table by projecting hill shading onto the slopes. A user interface element of a yellow sun depicted in 8 allows the user to change the direction of the sun and thus modify the hill shading.

Moreover, our platform provides a "flyto" feature that translates a laserpointer position on the sand table to latitude and longitude and to then fly to that location in Google Earth on a second screen. Fire scenarios that include historical progression maps can be loaded for post-mortem "lessons learned" review exercises (fig. 4).

There are important tradeoffs in different approaches to wildfire modeling. The biggest is the tradeoff of predictive accuracy for real-time feedback. As the initial uses for the Simtable have been for training and community outreach, we developed the CA approach as it allows users to get immediate interaction instead of waiting hours for a single fire to be simulated.

The Simtable is "modeling agnostic" in that it supports an application programming interface (API) to run user-supplied models that would take the elevation of the sand as an input. More sophisticated models can be loaded that add features like wind models that are terrain-sensitive and fire models that generate their own weather.

The platform can incorporate a variety of models, employing a number of underlying algorithms, as illustrated in the following section of this paper.

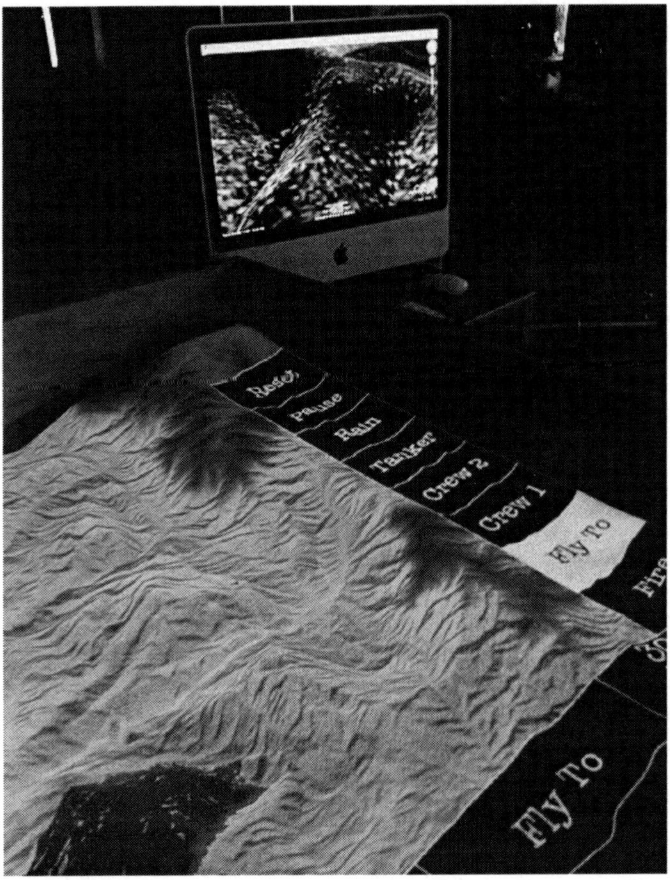

Figure 4: "Fly To" mode allows users to point to a location on the sand table and fly to that location on a separate first-person monitor, displaying the Google Earth terrain at the chosen location.

3 Forest fire applications on the platform

Our tangible interactive 3D platform can be used in a variety of contexts [1, 2]. The following sections illustrate how the platform is being applied in the real world for the training of firefighters and first responders, with the goal of improving the coordination among public safety agencies in the event of a major forest fire [3].

In this context, there are three main phenomena and activities that our platform allows to concurrently model for the training of first-responders and other emergency personnel in the containment and management of a wildfire:

1. The spreading of the wildfire
2. The evacuation dynamics
3. The firefighting strategy

3.1 Wildfire spreading model

Our platform can incorporate a variety of fire spreading models. The current application [3] trains wildland firefighters by simulating actual fire behavior, which spreads downwind, and uphill over mountain slopes, variably combusting vegetation fuel it finds in its path.

The table currently ships with a custom cellular automata (CA) model [3]. Fires can be interactively "started" on the sand table using either a physical cigarette lighter or a laserpointer. Ignited cells spread to neighboring cells in a probabilistic discrete event simulation. Fire spread rates are configurable as matrices in external .csv textfiles. Current matrices are draw from BehavePlus [5] for given temperatures, relative humidities, wind, slope and fuel types. When a fire ignites in a cell, the probability of a neighboring cell igniting depends on the wind, the elevation difference and the spread rate of the fuel type in that cell. Simply speaking, a cell is more likely to ignite if it is downwind and uphill from an already burning cell and it will combust more or less rapidly depending on its prevailing vegetation cover.

3.2 Evacuation dynamics

The wildfire evacuation model explores the interaction of two complex systems; a traffic model and a wildfire model. Combining two models with different time steps and architectures is a challenge in agent-based modeling, especially when the models are directly affecting each other. Cars blocking roads in an emergency evacuation could hinder fire crews preventing them from reaching the fire. This would cause the fire crew's attack on the fire to be hampered by the dynamics of the traffic model with a resulting change in fire model behavior as compared to an ideal fire attack (i.e. one with immediate fire crew presence). As the fires most likely to affect Santa Fe would be fast moving (up to 5 km/hour), we assume that only fires where the initial attack was unable to control the fire would rise to the level of evacuation. Thus this model can be considered the worst-case situation of no fire attack allowing the assumption that fire dynamics are independent of the traffic dynamics. This leaves the fire model free to be run independently of the traffic model and to serve as input to the traffic model. Below we detail the traffic model GIS inputs and traffic behavior including the wildfire model impact on the traffic model.

Initial explorations in modeling intersections were prototyped in NetLogo [11] with the road network modeled as a graph with nodes and edges. As an aside, our team continues to find NetLogo invaluable as a powerful rapid prototyping environment for agent-based models while it remains approachable to domain experts that don't normally identify themselves as coders. As the number of cars in the traffic model increased up to 70,000 and the GIS components grew in importance, development was transitioned to pure Java.

Agent-based models typically consist of agents interacting with each other in an environment. In general an environment may consist of cells in a grid with a topology of four or eight neighbors or a network (graph) topology. The traffic model's environment is constrained to the topology of the streets of Santa Fe as input in a GIS shapefile. The shapefile consists of shapes called polylines, shapes that are made up of a number of piecewise linear segments, and is a description of the 'center line' of the streets of Santa Fe. From this information a road network is built. Roads consist of multiple lanes proceeding from a directional distance of zero at one end to the total length of the road at the other. Each road has at least one lane that has directionality and a distance of zero at one end of the road proceeding to the length of the road at the other (fig. 5).

Depending on the number of lanes and whether the road is one-way, roads may have lanes beginning at both ends of the road. The cars exist on lanes that meet at intersections and can pass from one lane to another at intersections based on the connectivity of the intersection. Lanes that enter an intersection are connected to lanes that exit that intersection.

Car agents are aware of their neighbors sharing a lane, a road, or an intersection. Cars are not aware of their absolute positions in space such as elevation or north or east location as they exist at a distance along a lane in this world of roads and intersections. Cars are aware of their neighbors in this space, avoid collisions, accelerate and decelerate, and turn at intersections following a mixture of local rules while seeking a destination.

Predetermined sets of origins and destinations are defined in the model. At the beginning of a model run cars are assigned a destination and an origin location based on the real locations of homes in Santa Fe from another GIS data file. Destination points may include an evacuation center or road that leads out of town and can conceivably be extended to various other destinations, such as area hotels.

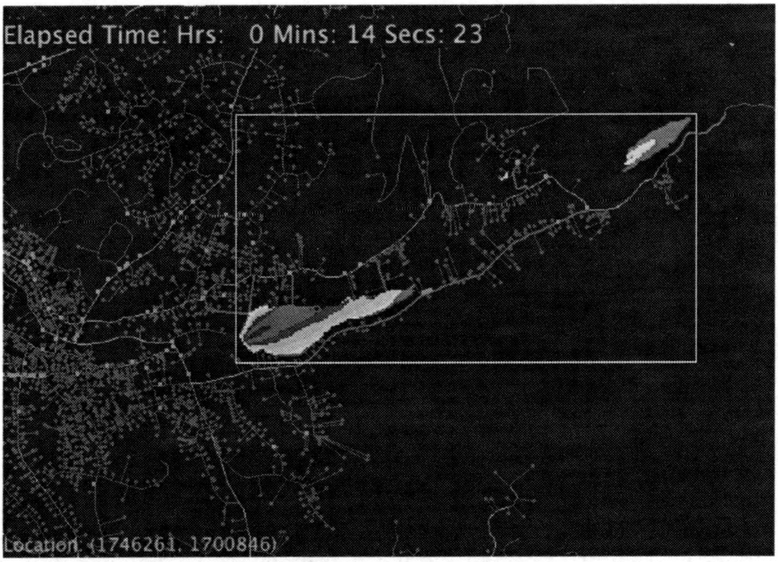

Figure 5: Agent-based traffic evacuation.

All intersections in the model are informed of the distance to destination points through an initial "*flood-fill*" from these points. This technique is useful when there are only a handful of possible destinations; however, it is memory and CPU-intensive for a large number of points.

This evacuation model is not so much concerned with the destination as with the dynamics of leaving the evacuated areas. Careful selection of evacuation points allows us to produce realistic sets of local routes out of the evacuated area. From that point, cars follow a set of local rules choosing particular routes at intersections according to the capacity of roads, their distance to the goal as determined by the flood fill, with aversion to changing roads too often, and a certain amount of random noise that we refer to as the 'tourist factor' that serves to give drivers an incomplete knowledge of the roads. Added to the above is general aversion to driving on or near roads that are near active fires. Possible further work in the social modeling of the evacuees would be necessary to determine where in particular populations would drive to, including friends and family homes, hotels, and other locations outside of the evacuated area. We focused, however, on areas which we knew would be challenging to evacuate in the event of an aggressive crown fire. The body of research produced by Cova *et al.* provides good background and many valuable techniques for characterizing relative neighborhood evacuation risk [6–10].

The fire model output consists of eight raster files of which the "Time of Arrival" (TOA) and "Flame Length" (FML) files are currently used as input to the traffic model. The TOA file consists of a geographic area covered by a grid of pixels. Pixel values are either the fractional hour that fire first enters that pixel or "–1" if no fire ever entered the area. Shape files generated from wildfire simulations were imported into Google Earth Pro (http://earth.google.com/) so that they could be shared with citizens of Santa Fe as .kmz layers. The Pro version of Google Earth is only required to generate the .kmz files from .shp files. The .kmz files are then distributable to end-users with the free version of Google Earth.

The traffic model takes this information as input and sorts fire points according to the time of the start of fire at each location. As the traffic model reaches the time when a fire first appears, that point is added to a set of current fires and roads that are near these fires are impacted as are roads that are located downwind of the fire. These points last for a certain time based on the estimate of duration of fire given the fuels at that location, before they are removed from the list. This information is made available to nearby roads, which can then be queried by cars traveling on these roads for deciding whether to turn at an intersection. Finding fire or smoke on a road, the car will avoid that road.

3.3 Firefighting strategies

The platform herein described is capable of not only simulating the spreading of fire and the evacuation of people, but it can also play out operational firefighting strategies that can be used to train fire crews.

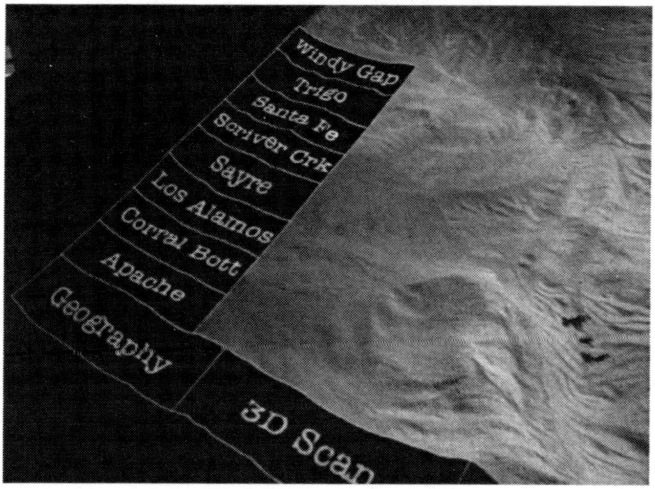

Figure 6: Historical fires menu.

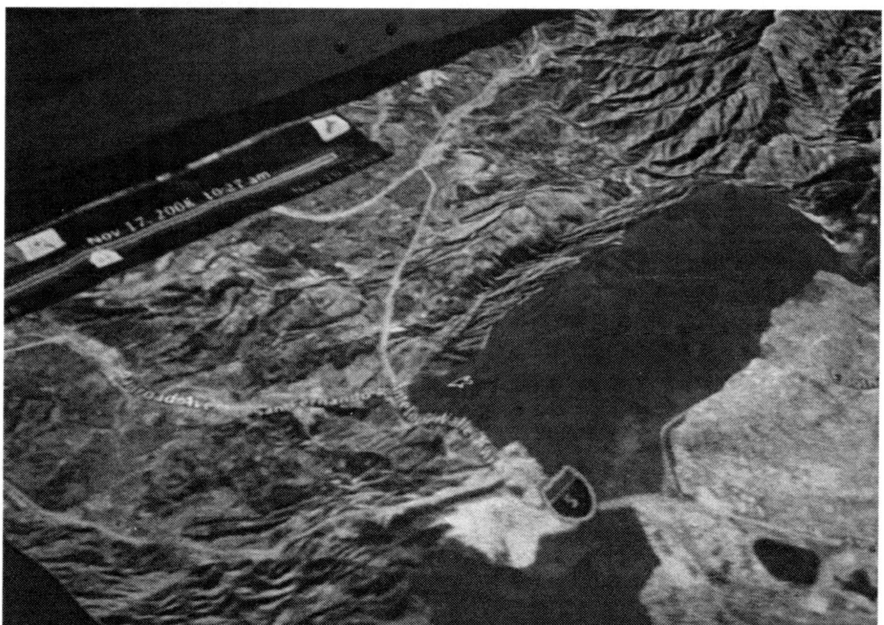

Figure 7: Historical fire progression example (Sayre Fire, 2008).

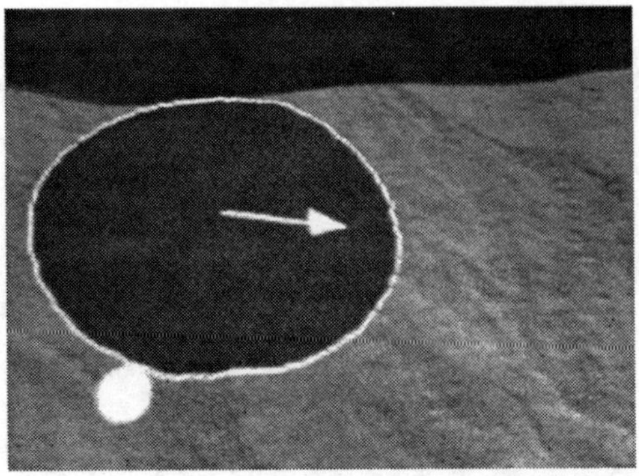

Figure 8: Interactive wind vector and sun location interface.

Figure 6 shows how our platform allows trainers to select eight "famous" wildfires and their corresponding topologies, to provide real-life scenarios on which firefighters can hone their skills.

Once a specific wildfire is selected, our platform displays a color-coded image of the terrain, suggesting where the users need to add or remove sand to conform it to the selected wildfire's terrain. The user can re-scan the surface to iteratively reproduce the terrain with the best approximation required for the training session (fig.7).

GIS layers, such as vegetation fuel, roads, and buildings can be interactively loaded on to the sand to affect fire behavior during a training session.

While the simulation is running, firefighter trainers are able to manipulate the strength and direction of the wind shown in fig. 8. Changing the wind, instantly affects the fire's behavior, thus requiring a quick reaction on the part of the trainees.

The latest version of the Simtable [3] supports the deployment of two types of crews (type 1 and type 2) who can clear fire lines along a fire's path to slow its spread and protect sensitive populated areas. The system simulates the production rate (in "chains per hour") of the two types of crews as they manually create the fire line. To make the strategies even more realistic the system also allows the use of flying "tankers" to airdrop fire retardants over specific locations again in an effort to slow the spreading of a wildfire.

All of these tools for the training of firefighters leverage the flexibility and adaptability of our platform.

4 Conclusions

The interactive tangible 3D platform can support a variety of models and interactions for the visualization of the spreading of forest fires, of the consequent evacuations of local inhabitants, and of a variety of strategies that can be employed to suppress the fires. It allows hands-on training of firefighters on historical fires and can simulate real firefighting actions, such as the clearing of fire lines by ground crews and the dropping of fire retardants using airplanes.

The decoupling of the interface from the underlying models allows flexibility for both the choice of interactivity (laserpointer, fiducials, hand gestures, face tracking, etc.) and the choice in sophistication of the fire spreading and traffic evacuation models and will also permit the inclusion of proprietary models or open-source tools developed in the community, thanks to a planned Application Programming Interface (API).

The realistic accuracy of the terrain model visualizations, including the fly-to views on Google Earth, as well as the tangible nature of the sand table – which had already extensively proven itself in traditional firefighting training – engage the users more profoundly and are likely to bring about measurable benefits in the quality of the training that the firefighters receive.

The platform can also be used to educate citizens about how to react when a fire hits closer to home. Being able to show one's own house on the sand table makes our platform a much more effective public outreach tool, which has many more applications above and beyond the modeling and management of forest fires.

References

[1] Thorp, J., Guerin, S., Wimberly, F., Rossbach, M., Densmore, O., Agar, M., Roberts, D., Santa Fe on fire: agent-based modeling of wildfire evacuation dynamics". *Proceedings of the Agent 2006 Conference on Social Agents: Results and Prospects*, Sallach, D.L., C.M. Macal, and M.J. North (eds.), Argonne National Laboratory and The University of Chicago: September 21-23, 2006.

[2] Autonomous urban agents: a Santa Fe approach to City Knowledge. Keynote presentation, *UCGIS Summer Assembly*, Santa Fe, June 22-23, 2009. http://www.slideshare.net/carreraf/ucgis-summer-09-final.

[3] Simtable LLC, www.simtable.com

[4] Ambient Pixel, www.ambientpixel.com

[5] Behave Plus (Fire.org,), http://fire.org/index.php?option=content&task=category&sectionid=2&id=7&Itemid=26.

[6] Cova, T.J., and Church, R.L., Modelling community evacuation vulnerability using GIS. *International Journal of Geographical Information Science*, 11(8), pp. 763-784. 1997.

[7] Cova, T.J., and Johnson, J.P. Microsimulation of neighborhood evacuations in the urban- wildland interface. *Environment and Planning A*, 34(12), pp. 2211-2229, 2002.

[8] Cova, T.J., and Johnson, J.P. A network flow model for lane-based evacuation routing. *Transportation Research Part A: Policy and Practice*, 37(7), pp. 579-604, 2003.

[9] Cova, T.J. Public safety in the urban-wildland interface: Should fire-prone communities have a maximum occupancy? *Natural Hazards Review*, 6(3), pp. 99-108, 2005.

[10] Cova, T.J., Dennison, P.E., Kim, T.H., and Moritz, M.A. Setting wildfire evacuation trigger-points using fire spread modeling and GIS. *Transactions in GIS*, 9(4), pp. 603-617, 2005.

[11] Wilensky, U. NetLogo. http://ccl.northwestern.edu/netlogo/. Center for Connected Learning and Computer-Based Modeling, Northwestern University. Evanston, IL, 1999.

Section 2
Air quality and health risk models

Numerical modelling of 2003 summer forest fire impacts on air quality over Portugal

A. I. Miranda[1], V. Martins[1], M. Schaap[2], R. San José[3], J. L. Perez[3], A. Monteiro[1], C. Borrego[1] & E. Sá[1]
[1]*CESAM & Department of Environment and Planning, University of Aveiro, Portugal*
[2]*TNO, Department of Air Quality and Climate, The Netherlands*
[3]*Universidad Politécnica de Madrid, Campus de Montegancedo, Spain*

Abstract

In 2003 Portugal faced the worst fire season ever recorded. The main purpose of this work is to evaluate the effects of the 2003 forest fires on the air quality, applying four numerical modelling systems (LOTOS-EUROS, MM5-CMAQ, WRF/chem and MM5-CHIMERE), and to compare their results with air quality data from several monitoring stations in Portugal. Forest fire emissions have been calculated taking into account the most suitable parameters for Portuguese forest/fire characteristics and the area burned by each forest fire. They were added to the anthropogenic and biogenic gridded emissions, according to the fire location and assuming a uniform fire spread and injection into the mixing layer. Simulations were performed during August 2003 regarding gaseous and particulate matter pollutants. To better evaluate the impact of forest fire emission on the air quality, a baseline simulation was performed, including the "conventional" emissions, along with a forest fire simulation, which also considered emissions from forest fires. Modelling hourly results, namely particulate matter (PM) and ozone (O_3) concentration, values have been compared to measurement data at several monitoring locations. In general, the different modelling systems show a good performance, which improves when forest fire emissions are considered, particularly for the PM concentrations. The influence of the forest fire emissions in O_3 formation is not so evident and needs more attention. The evaluation of the impact of forest fires on the air quality should be included in air quality assessment procedures, specifically in areas that

WIT Transactions on Ecology and the Environment, Vol 137, © 2010 WIT Press
www.witpress.com, ISSN 1743-3541 (on-line)
doi:10.2495/FIVA100071

are often affected by forest fires, such as south Europe, and air quality modelling systems can be important tools to achieve this goal.

Keywords: air quality modelling, forest fires, atmospheric emissions, particulate matter, ozone.

1 Introduction

Smoke has to be considered as one of the several disturbing effects of forest fires; it contains important amounts of carbon monoxide and dioxide (CO and CO_2), methane (CH_4), nitrogen oxides (NO_x), ammonia (NH_3), particulate matter (PM), particles with a mean diameter less than 2.5 μm (PM2.5) and particles with a mean diameter less than 10 μm (PM10), non-methane hydrocarbons (NMHC) and other chemical compounds. The effects of these emissions are felt at different levels: from the contribution to the greenhouse effect [1, 2] to the occurrence of local atmospheric pollution episodes, including high O_3 concentrations at medium distances from the emission sources [3]. Moreover, impacts on air quality and human health can be significant, as individuals and populations are exposed to hazardous air pollutants [4].

To understand and to evaluate forest fire effects on air quality, several factors should be analysed and comprehensively integrated, namely: fire progression, fire emissions, atmospheric flow, smoke dispersion and chemical transformation. There are several numerical modelling systems in development, some of them already available, aiming to estimate the dispersion of smoke from forest fires and their impact on the air quality. However, the majority of these systems do not include all the mentioned factors. Crucial in all systems is the quality of the forest fire emission estimates. Recently, quite a few works used remote sensing data to estimate emissions from wild land fires and to examine the impacts of specific fire events on regional and urban air quality [5, 6]. Alternatively, detailed information on burnt area, fuel loads, vegetation type, burning efficiency and emission factors can be used to estimate forest fire emissions, when available (*e.g.* [7]). Both approaches are nowadays an ongoing research topic.

The main purpose of this work is to evaluate the effect of forest fires emissions on the air quality applying four numerical modelling systems (LOTOS-EUROS, MM5-CMAQ, WRF/chem and MM5-CHIMERE) along a particular fire season, and to compare their results with air quality data from several monitoring stations in Portugal.

2 The modelling systems

The air quality modelling applications were performed using four different air quality modelling systems: MM5-CMAQ; MM5-CHIMERE; LOTOS-EUROS; and WRF/chem. All are 3D chemical transport models aimed to simulate air pollution in the lower troposphere.

Both CMAQ and CHIMERE were driven by the meteorological mesoscale model MM5. The MM5 model is a non-hydrostatic mesoscale meteorological model [8, 9] that is widely used around the world for meteorological research and also for operational meteorological use. It is capable of producing 3D wind, temperature, humidity and other important meteorological parameters and variables during simulations of several hours and days. It is based on a nested-grid primitive-equation model, which uses a terrain following sigma vertical coordinates. The LOTOS-EUROS model used as meteorological information data provided by the Free University of Berlin, which are based on a diagnostic meteorological model.

The CMAQ model [10] is a Comprehensive Air Quality Model which simulates the chemical transformations and the dispersion of the pollutants in a 3D domain. CMAQ model is structured in a full modular way. The different configurations should be consistent with those prepared for the MM5 meteorological simulations. Different applications of the MM5-CMAQ air quality modelling system have been performed during the last decade [11].

CHIMERE was specifically developed for simulating gas-phase chemistry, aerosol formation, transport and deposition at European and urban scales. The model simulates the concentration of 44 gaseous species and 6 aerosol chemical compounds. The gas-phase chemistry scheme, derived from the original complete mechanism MELCHIOR, has been extended to include sulphur aqueous chemistry, secondary organic chemistry and heterogeneous chemistry. The aerosol model accounts for both inorganic and organic species, of primary or secondary origin. MM5-CHIMERE has been applied and tested several times for the Portuguese conditions [12].

The LOTOS-EUROS model includes the O_3 chemistry using a modified Carbon-Bond Mechanism 4 (CBM4) mechanism. The model incorporates primary (combustion) particles, sea salt and secondary inorganic aerosols. In the vertical the model has 4 layers up to the 3,500 meters following the dynamic mixing layer approach. The LOTOS-EUROS system has been used in several applications showing good agreement between the observed and the modelled data [13, 14].

WRF/chem is an online multiscale air pollution prediction system based on the Weather Research and Forecasting (WRF) model, which is coupled with different chemical mechanisms. Biogenic and anthropogenic emissions, deposition, convective and turbulent chemical transport, photolysis, and advective chemical transport are all treated simultaneously with the meteorology. WRF/chem is fully consistent since all transport is done by the meteorological model with the same vertical and horizontal coordinates (no horizontal and vertical interpolation), the same physics parameterization for subgrid scale transport and no interpolation in time. It is capable to simulate chemistry and aerosols from cloud scales to regional scales. It includes different aerosol modules in model approach, sectional approach and mass only from GOCART modelling system. The photolysis packages are all coupled to aerosols and hydrometeors. It includes 4D-VAR chemical data assimilation. WRF/chem has been developed by NOAA with contributions from NCAR, PNNL, EPA, and university scientists [15].

3 Case study

Each summer season wild land forest fires burn a considerable area of south European landscape. Summer 2003 was one of the most severe fire seasons experienced during the last decades in Southern Europe and, due to persistent extreme fire conditions, Portugal suffered the worst forest fire season that the country has faced in the last 23 years, with a total area burned of almost 5 times the average [16]. There were reports of more than one thousand people needing medical assistance due to smoke intoxications, burns and wounds from forest fires in Portugal [16]. Satellite images, like the one shown in Figure 1, and air pollutants concentration values measured by the Portuguese air quality monitoring network, highlighted the impact of forest fire emissions [17].

Large forest fires (defined by the Portuguese Authorities as fires greater than 100 ha) are responsible for the majority of the area burned in Portugal. In 2003 the large fires burned 96% of total area burned besides representing approximately only 1% of the total occurrences. Forest fire emissions were estimated for every large forest fire occurred in this 2003 fire season, based on the following equation.

$$E_i = EF_i \times \beta \times B \times A \tag{1}$$

where: E_i – emission of compound i (g); EF_i – compound i emission factor (g.kg^{-1}); β – burning efficiency; B – fuel load (kg.m^{-2}); A – area burned (m^2).

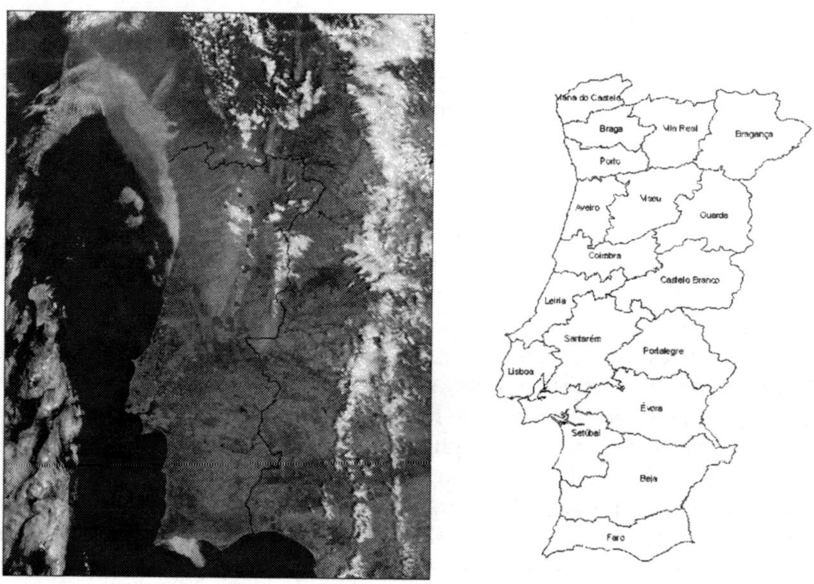

a) b)

Figure 1: Portuguese territory satellite image, 2003, August 3rd (a); Districts over Portugal (b).

Specific values for Portugal were selected based on data from the National Forest Inventory about the characteristics of the consumed forest type and shrubs. Furthermore, fire data like starting location and ignition time and area burned were collected from the National Forest Fires Inventory. The applied methodology has two components: (i) description of vegetation type, in terms of fuel load and combustion efficiency; and (ii) selection of the most adequate emission factors.

Forest fire emission values were added to the anthropogenic and biogenic gridded emissions, according to the fire location and assuming a uniform fire spread and injection into the mixing layer. Simulations with the four modelling systems were performed along August 2003 regarding gaseous and particulate matter pollutants. A baseline simulation (BS) was performed, including "conventional" emissions, and a forest fire simulation (FS), which also considered emissions from large forest fires.

The systems were firstly applied at the European scale and then over Portugal, using the same physics and a one-way nesting technique, and using the boundary conditions obtained from the coarser domain simulations. Table 1 lists the main characteristics of the air quality modelling systems applications.

For the European scale simulations, all systems used the emission inventory from the Netherlands Organisation for Applied Scientific Research (TNO), with the exception of CHIMERE that used the EMEP Program (Co-operative Programme for Monitoring and Evaluation of the Long-range Transmission of

Table 1: Main characteristics of the air quality modelling systems application.

Parameter	LOTOS-EUROS	MM5-CMAQ	WRF/chem	MM5-CHIMERE
Boundary conditions	Logan climatological datasets; LOTOS-EUROS European simulation (0.5° x 0.25° resolution)	MM5-CMAQ European simulation (50 km resolution)	WRF/chem European simulation (50 km resolution)	GOCART climatological models; MM5-CHIMERE European simulation (50x50 km^2 resolution)
Emissions	TNO inventory	TNO inventory	TNO inventory	EMEP and Portuguese inventory
Vertical structure	4 layers (up to 3,000 m)	23 layers (up to 10,000 Pa)	23 layers (up to 10,000 Pa)	8 layers (up to 3,500 m)
Chemical mechanism	CBM-IV	cb05-ae4	cbmz_mosaic_4bins	MELCHIOR
Horizontal grid resolution (km^2)	0.25° x 0.125°	16.6 x 16.6 km^2	16.6 x 16.6 km^2	10 x 10 km^2

Air Pollutants in Europe) emission inventory. For the Portuguese domain data from the most updated annual emission inventory (2003) developed by the Portuguese Agency for the Environment was used in case of the MM5-CHIMERE. Time disaggregation was obtained by the application of monthly, weekly and hourly profiles from the University of Stuttgart.

4 Results

Hourly modelling results were compared to monitored air quality data acquired at different background air quality stations. Air quality data were available at 13 of the 18 districts in Portugal (districts identification is depicted in Figure 1b). Most of the stations are located near the major conurbations at the west coast of the country, most notably Lisbon and Porto.

Some statistical parameters were estimated to better assess the simulation results, namely the root mean square error (RMSE), the systematic error (BIAS), and the correlation coefficient (r) [18]. Aiming to have a comparative picture of the modelling systems skills Figure 2 presents the RMSE, the BIAS and the r, for PM10 and O_3, averaged for each district, and considering the forest fire emissions.

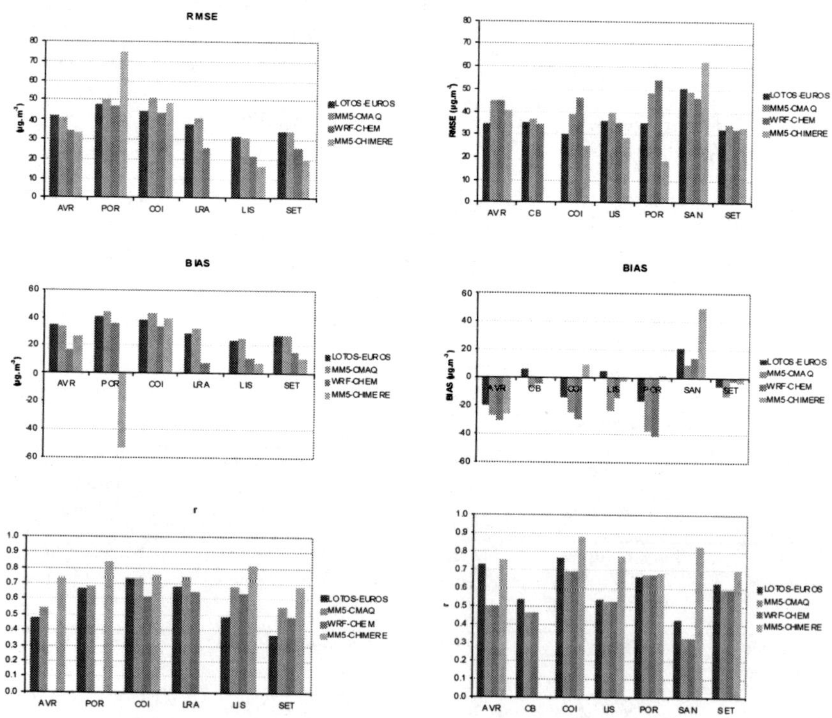

Figure 2: Averaged statistical indicators (RMSE, BIAS and r) concerning PM10 and O_3 for 2003, August, by district.

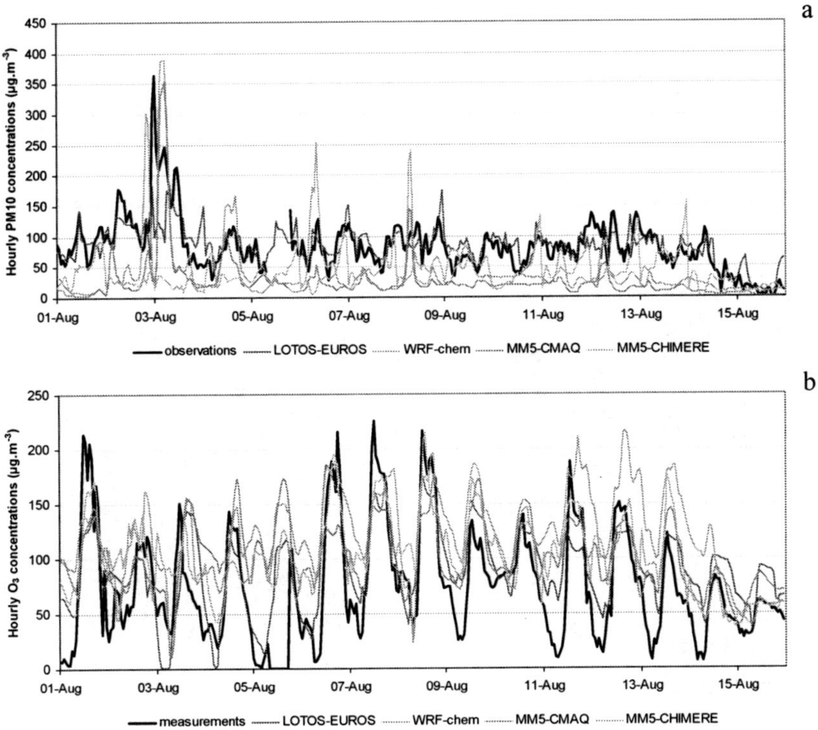

Figure 3: Hourly concentration values for PM10 (a) and O_3 (b), between the 1st and the 15th August 2003 at IGC station.

In general, there is not a significant difference among the models results. For PM10, the MM5-CHIMERE modelling system has low skills at Porto district results, with a higher RMSE and a negative BIAS, indicating an overestimation of PM10. That's the only overestimation for all the models and districts. Excluding Porto results, the RMSE varies between 15 and 50 $\mu g.m^{-3}$ and the BIAS between 5 and 40 $\mu g.m^{-3}$. The correlation coefficient is generally higher for MM5-CHIMERE and goes from 0.35 to 0.8.

Regarding O_3, the modelling systems tend to overestimate concentrations with a negative BIAS for almost all the districts and models. The RMSE vary between 20 and 60 $\mu g.m^{-3}$ and the correlation coefficient between 0.3 and 0.9.
Aiming to complement the statistical analysis Figure 3 shows the hourly time-series for PM10 and O_3 based on models results and measurements in a particular monitoring station - "Instituto Geofísico de Coimbra (IGC)". This monitoring station was selected because it is located in the central part of Portugal, which was one of the most affected ones by 2003 August forest fires. Data are presented for the first two weeks of August that were the most critical ones in terms of fire activity along this 2003 fire season.

Concerning PM10 and for the selected period, LOTOS-EUROS and MM5-CHIMERE systems tend to underestimate the measured peak PM10 values, namely at the 2nd and 3rd of August, when the fire activity was higher. MM5-CMAQ and WRF/chem were able to capture these particular peaks of PM10 that happened at the beginning of August. For the other days LOTOS-EUROS is the modelling system that better follows the measured values.

For O_3 the overestimation trend is confirmed. All the modelling systems tend to provide higher concentration values than the measured ones. Only when

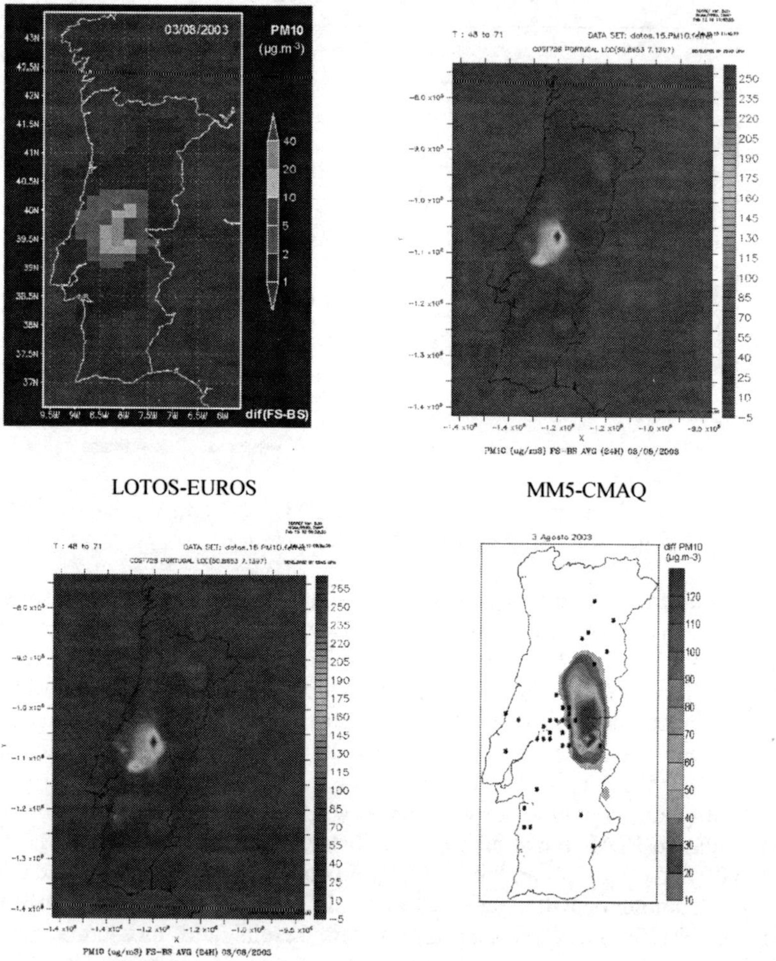

LOTOS-EUROS MM5-CMAQ

WRF/chem MM5-CHIMERE

Figure 4: Spatial differences (µg.m^{-3}) between simulation results with (FS) and without (BS) forest fire emissions, for PM10 daily averages on August 3rd, 2003.

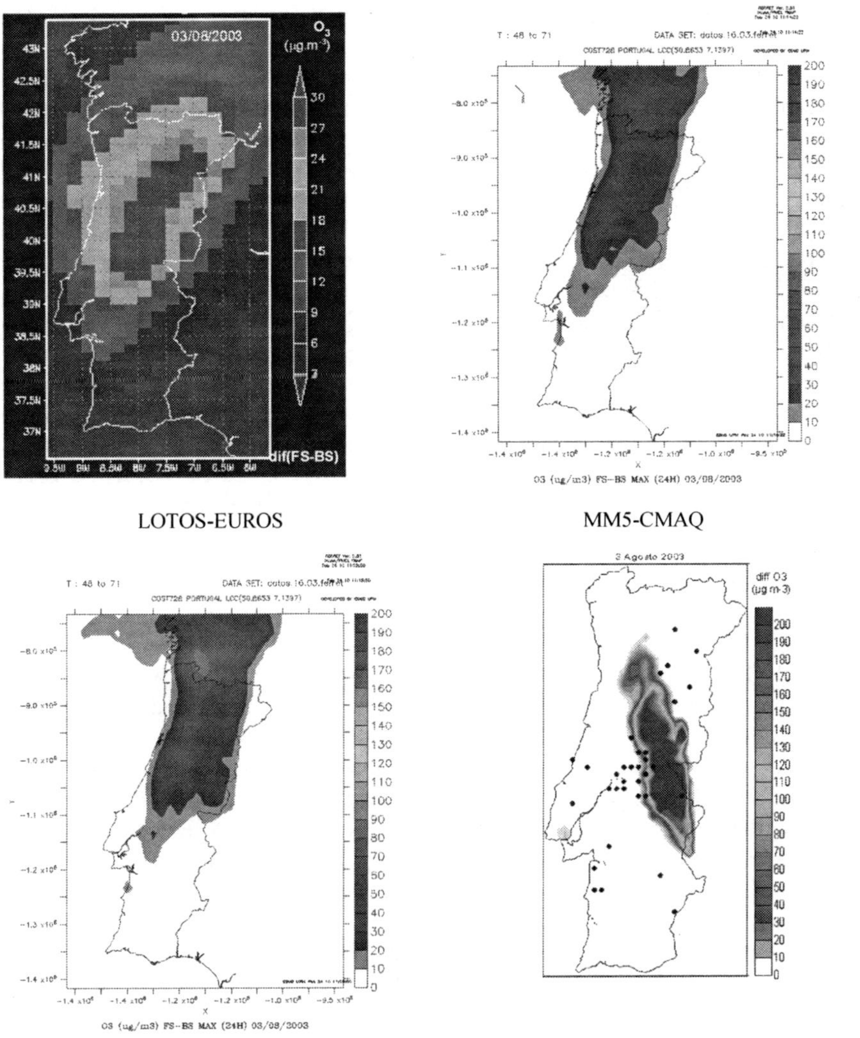

LOTOS-EUROS MM5-CMAQ

WRF/chem MM5-CHIMERE

Figure 5: Spatial differences (µg.m^{-3}) between simulation results with (FS) and without (BS) forest fire emissions, for O_3 daily maximum values on August 3rd, 2003.

measurements go upper than the 200 µg.m^{-3} (peak measured values) models were underestimating. The secondary character of ozone is clearly shown when comparing both pollutants series. Fires were spreading near Coimbra city, where the monitoring station is located, at the 3rd day of August and only PM10 measured values show their effects.

For this specific day, the impact of forest fires was higher at the central inland part of Portugal, and the PM10 daily mean difference reached 300 $\mu g.m^{-3}$, in case of MM5-CMAQ and WRF/chem. Comparatively, the models present different magnitude of the forest fires impacts, as consequence of run options (meteorological data, chemical mechanism, and others…). The slight deviation of the PM10 "plume" towards East regarding MM5-CHIMERE results could explain the difficulty of this modelling system to capture the measured peak shown in Figure 3.

The spatial analysis of results can contribute to a better understanding of the simulated values. Figures 4 and 5 show the spatial difference between both simulations (FS-BS) results, for this critical day (2003 August 3), concerning daily values for PM10 and maximum values for O_3, respectively.

The spatial differences for O_3 (Figure 5), as expected, are very different from the ones shown in Figure 4 for PM10. The forest fire emission affected a larger area of Portugal with the photochemical "plume" going to the North. MM5-CMAQ and WRF/chem even simulated a consumption of ozone in the central part of Portugal where fires were spreading and emitting nitrogen oxides.

5 Conclusions

This work investigated the impacts of forest fire emissions on the air quality over Portugal. The numerical modelling approach applied in this work confirms the significant impact of forest fire on atmospheric pollutants concentrations. In general, the different modelling systems show a good performance, which improves when forest fire emissions are considered, particularly for the PM10 concentrations. On the other hand, the influence of the forest fire emissions in O_3 formation is not evident and needs more attention. Future work will require that several questions should be analysed and integrated regarding the photochemical effects, namely the decrease in photolysis rates and increase in atmospheric radiative properties.

The evaluation of the impact of forest fires on the air quality should be included in air quality assessment procedures, specifically in areas that are often affected by forest fires as south Europe, and air quality modelling systems can be important tools to achieve this goal.

Acknowledgements

The authors thank the Portuguese Foundation for Science and Technology for the PhD grant of V. Martins (SFRH/BD/39799/2007) and for the Projects INTERFACE (POCI/AMB/60660/2004) and FUMEXP (PTDC/AMB/66707/2006) under the scope of the POCI2010 program and the European FEDER funds. In addition, COST 728 is acknowledged.

References

[1] Miranda, A.I., Coutinho, M., and Borrego, C. Forest fires emissions in Portugal: a contribution to global warming? *Environmental Pollution* 83, 121-123, 1994.

[2] Simmonds, P.G., Manning, A.J., Derwent, R.G., Ciais, P., Ramonet, M., Kazan, V. and Ryall, D. A burning question. Can recent growth rate anomalies in the greenhouse gases be attributed to large-scale biomass burning events? *Atmospheric Environment* 39, 2513-2517, 2005.

[3] Miranda, A.I., Borrego, C., Martins, H., Martins, V., Amorim, J. H., Valente, J. and, Carvalho, A. Forest Fire Emissions and Air Pollution in Southern Europe in: Earth Observation of Wildland Fires in Mediterranean Ecosystems. Springer. Berlin. 171-187, 2009.

[4] Coghlan B. The human health impact of the 2001-2002 “Black Christmas” bushfires in New South Wales, Australia: an alternative multidisciplinary strategy. *Journal of Rural and Remote Environmental Health*, 3(1), pp. 18 – 28, 2004.

[5] Hodzic, A., Madronich, S., Bohn, B., Massie, S., Menut, L., and Wiedinmyer, C. Wildfire particulate matter in Europe during summer 2003: meso-scale modelling of smoke emissions, transport and radiative effects. *Atmos. Chem. Phys. Discuss*., 7, 4705 – 4760, 2007.

[6] Sofiev, M., Vankevich, R., Lotjonen, M., Prank, M., Petukhov, V., Ermakova, T., Koskinen, J., Kukkonen J. An operational system for the assimilation of the satellite information on wild-land fires for the needs of air quality modelling and forecasting. *Atmos. Chem. Phys., 9, 6833-6847, 2009.*

[7] Ottmar, R., Miranda, A.I., Sandberg, D. Characterizing Sources of Emissions from Wildland Fires. In Wild land fires and air pollution. Developments in Environmental Science, Vol 8, Chapter 3. Elsevier B.V.: A. Bytnerowicz, M. Arbaugh, A. Riebau and C. Andersen, p.101-136, 2009.

[8] Dudhia J. A nonhydrostatic version of the Penn State / NCAR mesoscale model: Validation tests and simulations of an Atlantic cyclone and cold front. *Mon Weather Rev*, 121, 1493–513, 1993.

[9] Grell GA, Dudhia J, Stauffer DR. A description of the fifth-generation Pennstate/NCAR mesoscale model (MM5). NCAR/TN- 398+ STR. NCAR Technical Note, 1994.

[10] Byun, DW, Young, J, Gipson, G, Godowitch, J, Binkowski, F, Roselle, S, *et al.* Description of the Model-3 Community Multiscale Air Quality (CMAQ) model. Proceedings of the American Meteorological Society 78th Annual Meeting, Phoenix, AZ; 1998. p. 264–8. Jan 11–16, 1998.

[11] San José R, Pérez JL, González RM. A mesoscale study of the impact of industrial emissions by using the MM5-CMAQ modelling system. *International Journal of Environment and Pollution*. 22 (1/2), 144-162, 2004.

[12] Monteiro A., Miranda A.I., Borrego C., Vautard R., Ferreira J., Perez A.T. Long-term assessment of particulate matter using CHIMERE model.. *Atmospheric Environment. 41, 36, 7726-7738,* 2007.
[13] Schaap, M., van Loon, M., ten Brink, H.M., Dentener, F.D., Builtjes, P. Secondary inorganic aerosol simulations for Europe with special attention to nitrate, *Atmos. Phys. Chem.*, 4, 857-874, 2004.
[14] Schaap, M., Timmermans, R.M.A., Sauter, F.J., Roemer, M., Velders, G.J.M., Boersen, G.A.C., Beck J.P., Builtjes, P. The LOTOS-EUROS model: description, validation and latest developments. *International Journal of Environment and Pollution*, 32(2), 270-290, 2008.
[15] Grell, G., Peckham, R., Schmitz, S., McKeen, G., Frost, W., Skamarock, B. Fully coupled online chemistry within the WRF model, *Atmos. Environ.*, 39, 6957-6975, 2005.
[16] EC – European Commission. San-Miguel-Ayanz, J., Barbosa, P., Camia, A., Kucera, J., and Libertà, G. (Eds), Forest Fires in Europe - 2003 fire campaign -, Official Publication of the European Communities, SPI.04.142.EN, 2004.
[17] Miranda, A.I., Monteiro, A., Martins, V.; Carvalho, A., Schaap, M., Builtjes, P., Borrego, C. Forest fires impact on air quality over Portugal. In Air Pollution Modeling and Its Application XIX, C. Borrego and A.I. Miranda: Dordrecht: Springer, 2008.
[18] Borrego C., Monteiro A., Ferreira J., Miranda A. I., Costa A.M., Carvalho A. C., Lopes M. Procedures for estimation of modelling uncertainty in air quality assessment. *Environment International. 34, 613-620,* 2008.

Monitoring fire-fighters' smoke exposure and related health effects during Gestosa experimental fires

A. I. Miranda[1], V. Martins[1], P. Cascão[1], J. H. Amorim[1], J. Valente[1], R. Tavares[1], O. Tchepel[1], C. Borrego[1], C. R. Cordeiro[2], A. J. Ferreira[2], D. X. Viegas[3], L. M. Ribeiro[3] & L. P. Pita[3]
[1]*CESAM & Department of Environment and Planning, University of Aveiro, Portugal*
[2]*Centre of Pulmonology of Coimbra University Medical School, Portugal*
[3]*Association for the Development of Industrial Aerodynamics, University of Coimbra, Portugal*

Abstract

The main objective of this study is to contribute to the scientific knowledge regarding fire-fighters' exposure to smoke and its related health effects. Forest fire experiments were developed with an extensive number of measurements of individual exposure to smoke pollutants and of medical parameters for a group of fire-fighters. For the smoke exposure monitoring, ten fire-fighters from four different fire brigades were selected. The fire-fighters' individual exposure to toxic gases and particulate matter was monitored with portable devices, and their location in time was registered with GPS equipment. For all the monitored fire-fighters, air pollutant concentration values acquired during the fire experiments were beyond the limits recommended by the World Health Organization (WHO), namely for PM2.5, CO and NO_2. Daily averages of PM2.5 concentration values as high as 738 $\mu g.m^{-3}$ were obtained, well above the recommended limit of 25 $\mu g.m^{-3}$. In terms of CO, hourly averaged values higher than 73,000 $\mu g.m^{-3}$ were monitored, clearly above the 30,000 $\mu g.m^{-3}$ recommended by the WHO. The highest NO_2 hourly averaged measured value was 4,571 $\mu g.m^{-3}$, once again much higher than the recommended value of 200 $\mu g.m^{-3}$. For VOCs, a maximum hourly average of 10,342 $\mu g.m^{-3}$ was registered for one of the fire-fighters; however, due to the lack of recommended

WIT Transactions on Ecology and the Environment, Vol 137, © 2010 WIT Press
www.witpress.com, ISSN 1743-3541 (on-line)
doi:10.2495/FIVA100081

or legislated values it is not possible to establish a comparison. The medical tests conducted on the fire-fighters, before and after the exposure to smoke, also indicate a considerable effect on the measured medical parameters, in particular an expressive increase of CO concentration and a decrease of NO concentration in the exhaled air of the majority of the fire-fighters.

Keywords: smoke exposure, fire experiments, carbon monoxide, particulate matter, nitrogen dioxide, medical tests.

1 Introduction

Nowadays there is a growing awareness that smoke produced during forest fires can expose individuals and populations to hazardous concentrations of air pollutants. However, the current state of knowledge about the potential health impacts on fire-fighting personnel is still scarce, in particular within the European context. The most extensive measurements of smoke exposure among wild land fire-fighters were conducted in the United States of America (USA) and Australia [1–5]. From these field studies it was possible to conclude that fire-fighters can be exposed to significant levels of carbon monoxide (CO) and respiratory irritants, including formaldehyde, acrolein, and respirable particles [3, 5]. As a result, adverse health effects occur with acute, instantaneous eye and respiratory irritation and shortness of breath, developing into headaches, dizziness and nausea enduring for up to several hours. Additionally, long-term health effects, such as impaired respiratory function or increased risk of cancer, may be caused by these pollutants. Special concern is raised by exposure to respirable particles and potentially toxic compounds adsorbed to them (e.g. polycyclic aromatic hydrocarbons (PAHs) and semivolatile organic compounds, some of which may be carcinogenic), as well as to aldehydes, compounds that are known as probable human carcinogens. There are a number of factors that affect the impacts of smoke on health, including the concentration of air pollutants within the breathing zone of the fire-fighter, the exposure duration, exertion levels, and individual susceptibility, such as pre-existing lung or heart diseases [6].

In Europe, where an average annual value of 500,000 hectares of forest was consumed by fire in the last 29 years [7], there is a considerable lack of data on personal smoke exposure. These data are of vital importance for the establishment of cause/effect relationships between the exposure to air pollutants from smoke and fire-fighters' health effects.

Exposure results from the experiments in the USA and Australia may not be applicable to European wild land fire-fighters due to differences in vegetation, fire conditions and fire-fighting operations. The composition of smoke depends on the type of vegetation being burned, fuel moisture content, temperature of the fire and wind conditions [6]. Additionally, a major factor influencing exposure is the type of work activities that the fire-fighters carry out. Therefore it is crucial to assess exposure at the individual level and within the European context to determine whether exposure could result in health damage and what primary factors influence exposure.

The main purpose of this paper is to contribute to the fire-fighters' smoke exposure and related health effects knowledge. The current work presents and analyzes the data on individual exposure to CO, nitrogen dioxide (NO_2), volatile organic compounds (VOCs), and particles with an aerodynamic diameter lesser than 2.5 µm (PM2.5), which were obtained during field burning experiments for a group of ten fire-fighters equipped with portable "in continuum" measuring devices. A group of 14 fire-fighters were also tested before and after fire-fighting regarding their exhaled nitric oxide (eNO) and CO.

2 Methodology and equipment

The measurement of fire-fighters individual exposure and of the medical parameters was conducted during the Gestosa 2008 and 2009 fire experiments, in Central Portugal, at the end of spring season.

2.1 Study area characteristics

The study area is located in the mountain range of Lousã, Central Portugal, at an altitude between 900 and 1,100 m. The vegetation was mainly composed by continuous shrubs of three dominant species: *Erica umbellata*, *Ulex minor* and *Chamaespartium tridentatum*. In Figure 1 it is possible to have a perspective of the study area general characteristics, for 2008 and 2009 fire experiments.

The study areas were divided into 7 and 4 plots in 2008 and 2009, respectively, with regular shapes and variable dimensions. For 2008 plots varied between 874 and 2,820 m^2 and for 2009 plots varied between 1,800 and 6,057 m^2. These experimental burning plots were established within Forest Service lands, and within the Gestosa forestry perimeter.

Before the experiments the burning plots were prepared and the vegetation properties analysed. The characteristics of the experimental plots and available fuel are presented in Table 1.

During one month before the experiments, hourly data related with wind speed, wind direction, precipitation, air temperature and relative humidity were

Figure 1: Plot layouts from the Gestosa 2008 and Gestosa 2009 study areas.

Table 1: Main characteristics of the experimental plots (Gestosa 2008 and 2009).

Plot	Area (m^2)	Slope (°)	Fuel cover (%)	Fuel height (cm)	Fuel bulk density ($kg.m^{-3}$)	Fuel load ($ton.ha^{-1}$)
Gestosa 2008						
P01	820	20	100.00	83.25	2.04	24.79
P02	959	27	100.00	93.00	2.06	26.69
P03	1,228	24	98.20	85.95	2.11	26.31
P04	1,493	22	86.00	70.40	2.26	22.41
P05	2,642	20	100.00	66.53	2.23	33.58
P06	1,089	23	100.00	83.00	2.28	31.17
P07	1,049	17	100.00	66.25	2.34	29.15
Gestosa 2009						
P11	2,552	19	*	*	*	*
P12	1,800	17	*	*	*	*
P13	6,057	14	*	*	*	*
P14	2,990	19	*	*	*	*

* Plots are safety areas with little vegetation.

recorded by a Geolog S meteorological station. This information allowed assessing the best period of the day to burn with the advisable wind conditions.

The duration of the burns in Gestosa 2008 was rather small (10-15 minutes) when compared to wildfires. Although the fire duration during Gestosa 2009, which lasted for almost one hour for a specific plot, is higher than during Gestosa 2008, the plots had little vegetation and that limited the fire-fighter's exposure to smoke.

2.2 Smoke exposure

For the smoke exposure monitoring, 10 fire-fighters were selected from four different fire corporations. Fire-fighters were chosen based on predefined criteria that took into account the age, smoking habits, respiratory diseases and function in the fire brigade. The selected fire-fighters were equipped with sampling devices monitoring individual exposure to CO, VOC, NO_2 and PM2.5. Moreover, the location of each corporation in time was registered with GPS equipment. For the selection of the monitoring equipment some important aspects were considered, namely their weight and the robustness, as well as the measuring ranges. Figure 2 shows some fire-fighters with the exposure monitoring equipment.

VOC and NO_2 were monitored continuously using integrated photo-ionization detector GasAlertMicro 5 PID from BW Technologies. The rechargeable battery allows a continuous operation up to 12 hours and with the memory card is capable of recording two months of data. The VOC and NO_2 sensors were calibrated before the burn using a 100 ppm isobutylene and 10 ppm NO_2 calibration gas, respectively.

Figure 2: Fire-fighters with the exposure monitoring equipment.

Table 2: Characteristics of the equipment.

Pollutant	Type of data	Equipment	Characteristics	
			Range	Resolution
VOC	Continuous measurement: 5 seconds interval	GasAlertMicro 5 PID from BW Technologies	0-1,000 ppm	1 ppm
NO_2			0-99.9 ppm	0.1 ppm
CO	Continuous measurement: 5 seconds interval	GasAlertMicroClip from BW Technologies	0-500 ppm	1 ppm
		GasAlertextreme from BW Technologies	0-1,000 ppm	1 ppm
PM2.5	Continuous measurement: 1 minute interval	Personal Aerosol Monitor SidePack AM510 from TSI	0-20 $mg.m^{-3}$	0.001 $mg.m^{-3}$

CO was also monitored continuously using a CO GasAlertMicroClip and CO GasAlertextreme from BW Technologies, in Gestosa 2008 and 2009, respectively. The CO detector was calibrated before the fire experiments using a 100 ppm CO calibration gas.

PM2.5 monitoring was performed using the monitor SidePack AM510 Personal Aerosol Monitor from TSI Inc. fitted with a built in 2.5 μm cut off impactor at a constant flow rate of 1.7 $L.min^{-1}$. Before the fire experiments the flow rate was calibrated and the monitor was zeroed using a zero filter. Table 2 summarizes the characteristics of the equipments.

2.3 Air quality limit values

Aiming to better understand the effects of these experimental fires on the fire-fighters health, the measured results were compared to the European air quality legislation values and to the values recommended by the WHO (Table 3).

Both the air quality legislation limit values as well as the WHO standards were established aiming to protect the human health from the air pollution effects. In general, the proposed limit values agree with only one exception regarding PM2.5, for which WHO is more demanding. Moreover, the WHO recommends limit values for other time periods than those considered by the European Directive.

2.4 Medical tests

The respiratory function of a 38 fire-fighters sample was evaluated, prior to any exposure, during April 2008. They also answered the SF-36 questionnaire, which regards the general quality of health. An initial subgroup of 14 non smoker fire-fighters was tested during 2008, before and after fire-fighting, regarding to their eNO and CO. During the Gestosa 2009 experiments, eNO, CO and % carboxy-haemoglobin (COHb) were also registered for a similar sub-group of 14 fire-fighters, pre and post smoke exposure. In 2009, a sample of exhaled breath condensate was collected too, before and after smoke exposure, for the determination of lung inflammatory patterns. Table 4 summarizes the characteristics of the medical equipments.

Table 3: Air quality limit values for the protection of human health established by European legislation and recommended by the WHO.

Pollutant	European Legislation (2008/50/CE)	WHO
PM2.5	25 $\mu g.m^{-3}$ (1 year)	25 $\mu g.m^{-3}$ (24 hour)
		10 $\mu g.m^{-3}$ (1 year)
NO_2	200 $\mu g.m^{-3}$ (1 hour)	200 $\mu g.m^{-3}$ (1 hour)
	40 $\mu g.m^{-3}$ (1 year)	40 $\mu g.m^{-3}$ (1 year)
CO	10 $mg.m^{-3}$ (8 hours)	100 $mg.m^{-3}$ (15 minutes)
		60 $mg.m^{-3}$ (30 minutes)
		30 $mg.m^{-3}$ (1 hour)
		10 $mg.m^{-3}$ (8 hours)

Table 4: Characteristics of the medical equipment.

Parameter	Equipment	Characteristics	
		Range	Resolution
eNO	Nioxmino from Aerocrine	5-300 ppb	1 ppb
Alveolar CO	MICRO CO/Smoke-check from Micromedical	0-500 ppm	1 ppm

3 Results and discussion

3.1 Smoke exposure

To assess fire-fighters exposure to smoke pollutants and taking into account the recommended limit values (see Table 3), 1 hour averages for CO, VOC and NO_2 and 24 hours averages for PM2.5 were calculated for every monitored fire-fighter. Table 5 presents the maximum hourly-averaged values for CO, NO_2 and VOC and the daily averages for PM2.5. This last one was calculated considering zero PM2.5 values for the non-exposure time periods.

For all the monitored fire-fighters, a considerable number of the air pollutants concentration values acquired during their activity is beyond the limits recommended by the WHO (see bold values in Table 5), namely for PM2.5, CO

Table 5: Highest hourly averages of CO, NO_2 and VOC and 24 hour averages for PM2.5.

Fire-fighter	CO	NO_2	VOC	PM2.5
	($\mu g.m^{-3}.h^{-1}$)			($\mu g.m^{-3}.day^{-1}$)
Gestosa 2008				
1	**32,479**	**2,163**	1,585	**260**
2	**73,033**	**4,172**	3,934	**184**
3	**47,223**	**3,641**	415	**306**
4	**50,881**	**274**	1,789	**240**
5	**33,178**	**709**	599	**738**
6	**49,078**	n.d.	2,917	**735**
7	n.d.	**2,599**	1,838	**684**
8	**35,847**	**609**	1,520	**479**
9	**48,259**	**4,571**	5,302	**610**
10	n.d.	82	2,097	**206**
Gestosa 2009				
1	12,586	**344**	54	**44**
2	22,814	**332**	526	**400**
3	**32,222**	**485**	496	**124**
4	**39,090**	142	343	**315**
5	**36,199**	**884**	337	**152**
6	**30,669**	**1,544**	10,342	**40**
7	4,903	132	62	**66**
8	**41,9389**	**788**	1,377	**396**
9	**42,023**	**802**	376	**176**
10	17,899	**1,091**	1,076	**358**

n.d. – No data

and NO_2. Daily averages of PM2.5 concentration values as high as 738 μg.m^{-3} were obtained, well above the recommended limit of 25 μg.m^{-3}, even considering that during the rest of the day the concentration was 0 μg.m^{-3}. In terms of CO, hourly averaged values higher than 73,000 μg.m^{-3} were monitored, clearly above the 30,000 μg.m^{-3} recommended by the WHO. The highest NO_2 hourly averaged measured value was 4,571 μg.m^{-3}, once again much higher than the recommended value of 200 μg.m^{-3}. For VOC, a maximum hourly average of 10,342 μg.m^{-3} was registered for one of the fire-fighters; however, due to the lack of recommended or legislated values for total VOC it is not possible to establish a comparison.

Aiming to have the time evolution of exposure values along the experiments, hourly averaged values for the measured pollutants during Gestosa 2008 and Gestosa 2009, for two particular fire-fighters, are presented in Figure 3.

The hourly averages for the pollutants show a similar pattern in terms of concentration variation, which is related to the smoke exposure. PM2.5 and CO are the pollutants that present the highest concentrations.

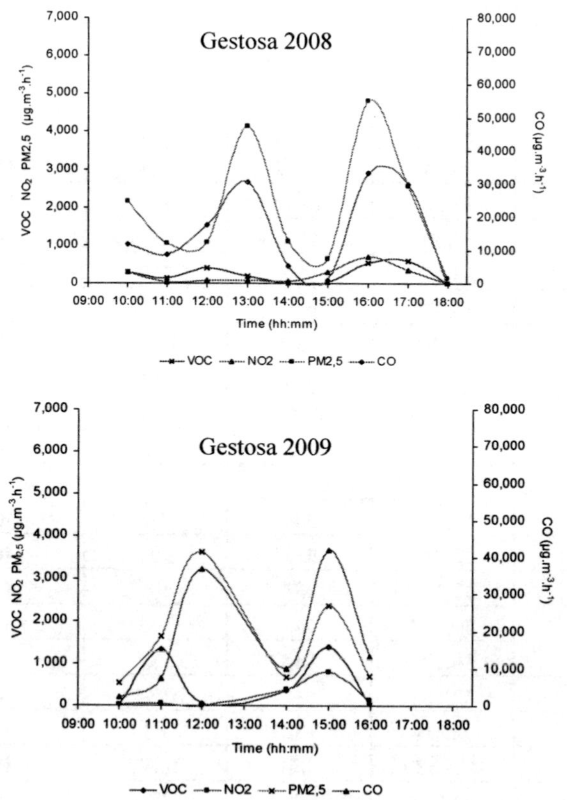

Figure 3: Hourly averaged exposure values in Gestosa 2008 and 2009 for fire-fighters 5 and 8.

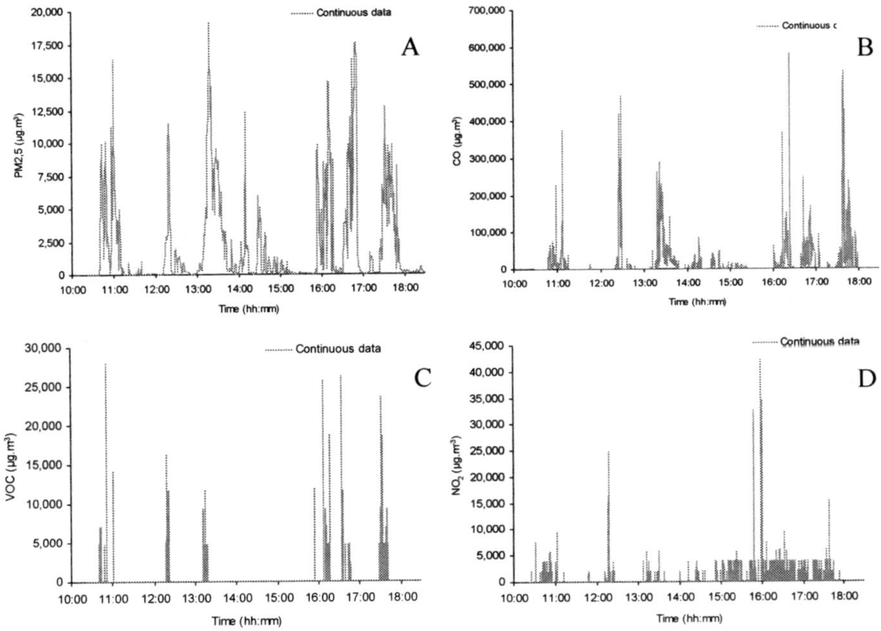

Figure 4: PM2.5, CO, VOC and NO_2 concentrations measured during Gestosa 2008 for fire-fighter 5 (A, B, C and D respectively).

Figure 4 also shows the instantaneous registered data along the Gestosa 2008 experiments, for the fire-fighter 5.

The instantaneous CO concentration values acquired during the Gestosa 2008 were very high, reaching a maximum value above 600,000 $\mu g.m^{-3}$. PM2.5 values were also very high (19,953 $\mu g.m^{-3}$). These data show the magnitude of the exposure peaks occurred during regular fire-fighting operations. For instance, the knowledge of the CO concentration peaks to which fire-fighters are exposed is quite important, since high concentrations of this gas can cause death by asphyxia. The same type of results was obtained for Gestosa 2009.

3.2 Health assessment

Figures 5 and 6 illustrate the changes in the medical measured parameters in Gestosa 2008 and 2009, respectively.

The medical tests conducted on the fire-fighters in 2008, before and after the exposure to smoke, indicate a considerable effect on the measured parameters. Regarding CO concentration in the exhaled air: (i) there was a higher number of fire-fighters with concentration values above 7 ppm after fire (20 ppm were even registered for one fire-fighter); (ii) before fire 11 fire-fighters had CO levels in the [0-6] ppm range, and after fire only 2 remained in the same interval.

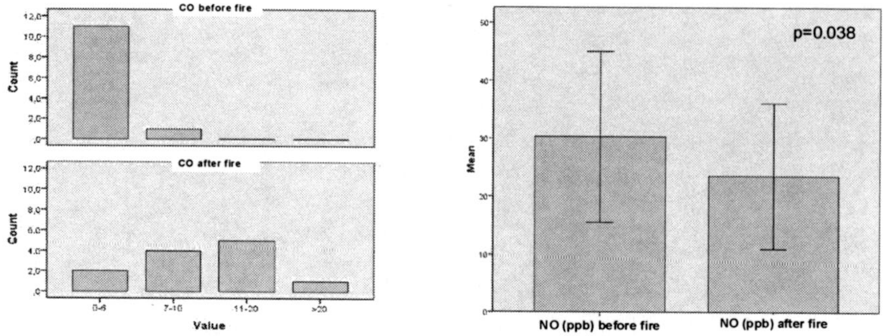

Figure 5: Medical test results for Gestosa 2008 before and after fire exposure, for CO (ppm) and NO (ppb).

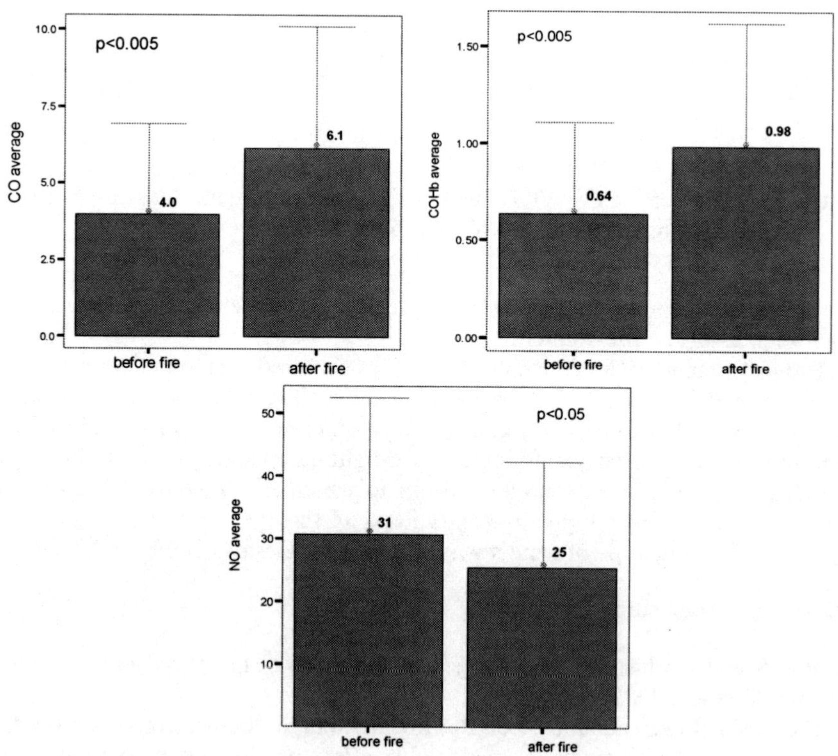

Figure 6: Medical tests results for Gestosa 2009 before and after fire exposure (CO and COHb in ppm, NO in ppb).

In what concerns the exhaled NO, there was a significant decrease ($p=0.038$) between the values measured before and after the exposure to smoke, achieving 5 ppb. This could indicate an effect similar to the exposure to cigarette smoke. Indeed, in current smokers, it is usually observed a decrease on exhaled nitric oxide, probably related to the inhibition of nitric oxide synthetase [8].

Figure 6 shows, for CO, COHb and NO, the averaged values for the 14 monitored fire-fighters. In the scope of the previously observed, there was a strong increase ($p<0.005$) of the exhaled CO and COHb after fire, reaching 2.10 ppm and 0.34 ppm, respectively. In terms of NO in the exhaled air, there was a decrease ($p<0.05$) in the order of 6 ppb.

4 Conclusions

Usually, the amount and characteristics of noxious exposure of forest fire-fighters are not widely recognized; more attention has been drawn upon the risks of indoor fire-fighting. Our work indicates that forest fire-fighting can expose individuals to very high concentrations of CO, VOC, NO_2 and PM2.5, with potential harmful effects on human health. Urgent measures to avoid these levels of exposure are needed. They can be related to the use of adequate protecting devices, to a correct planning of fire-fighting shifts, and/or to the operational availability of information regarding the areas of higher pollutants levels that can be obtained through modelling of exposure.

Acknowledgements

The authors would like to acknowledge the financial support of the Portuguese Ministry of Science, Technology and Higher Education, through the Foundation for Science and Technology (FCT), for the PhD grants of V. Martins (SFRH/BD/39799/2007), J. Valente (SFRH/BD/22687/2005), R. Tavares (SFRH/BD/22741/2005) and the Post-Doc grant of J.H. Amorim (SFRH/BPD/48121/2008). In addition, FCT is acknowledged for the funding of the National research project FUMEXP (PTDC/AMB/66707/2006) through the POCI2010 program and the FEDER fund.

References

[1] McMahon, C.K., Bush, P.B., Forest worker exposure to airborne herbicide residues in smoke from prescribed fires in the southern United-States. *Am Ind Hyg Assoc J* 53(4), pp. 265–72, 1992.

[2] Materna, B.L., Koshland, C.P., Harrison, R.J., Carbon monoxide exposure in wildland firefighting: a comparison of monitoring methods. *Appl Occup Environ Hyg*, 8(5), pp.479–87, 1993.

[3] Reinhardt, T.E., Ottmar, R.D., Smoke exposure at western wildfires. USDA Forest Service Pacific Northwest Research Station Research Paper (525), 2000.

[4] Reinhardt, TE, Ottmar, RD, Hanneman, A., Smoke exposure among firefighters at prescribed burns in the Pacific Northwest. USDA Forest Service Pacific Northwest Research Station Research Paper (526), pp. U1–45, 2000.

[5] Reinhardt, T.E., Ottmar, R.D., Baseline measurements of smoke exposure among wildland firefighters, *J Occup Environ Hyg* 1(9), pp. 593–606, 2004.

[6] Reisen, F., Brown, S.K., Australian firefighters' exposure levels to air toxics during bushfire burn of autumn 2005 and 2006. *Environment International* 35, pp. 342-352, 2009.

[7] European Commission, Forest Fires in Europe 2008. EUR 23971 EN. ISSN 1018-5593, 2009.

[8] Malinovschi, A., Janson, C., Holmkvist, T., Norback, D., Merilainen, P., M. Hogman, N., Effect of smoking on exhaled nitric oxide and flow independent nitric oxide exchange parameters. *Eur Respir J* (28), pp. 339–345, 2006.

Section 3
Detection, monitoring and response systems

An integrated approach for early forest fire detection and verification using optical smoke, gas and microwave sensors

N. von Wahl[1], S. Heinen[1], H. Essen[1], W. Kruell[2],
R. Tobera[2] & I. Willms[2]
[1]*Fraunhofer Institute for High Frequency Physics and Radar Techniques FHR, Germany*
[2]*Department of Communication Systems (NTS), University of Duisburg-Essen, Germany*

Abstract

In 2008 the research project "International Forest Fire Fighting" (iWBB) was funded by the Minister for Economic Affairs and Energy of the State of North Rhine-Westphalia, Germany. A group of companies, research institutes and universities are working together to develop an integrated, but modular system. An integrated approach for early forest fire detection and suppression is based on an adequate combination of different detection systems depending on wildfire risk, the size of the area and human presence affiliated with an adequate logistical infrastructure, training by simulation, and innovative extinguishing technology based on armoured tracked fire fighting vehicles.

As large areas have to be monitored, only remote sensing technologies (e.g. video based systems) are able to perform early detection adequately. To reduce false alarms a remote controlled unmanned aerial vehicle (UAV) equipped with gas sensors and a thermal camera flies to a potential fire to specify the origin of the reported cloud. The UAV can also be used as a scout for fire fighters. After successful fire extinction an unmanned blimp can be used as a fireguard to reduce the risk of reignition of the fire. As monitoring tools, a microwave radiometer, which is capable of detecting hot spots even under insufficient vision (due to smoke clouds and below the ground surface), gas and smoke sensors and a thermal camera are mounted on the blimp. The benefit of a blimp is a higher payload.

WIT Transactions on Ecology and the Environment, Vol 137, © 2010 WIT Press
www.witpress.com, ISSN 1743-3541 (on-line)
doi:10.2495/FIVA100091

This paper presents an investigation of an early forest fire detection system on the basis of indoor (performed in the fire lab of the University of Duisburg-Essen) and outdoor tests. A commercial highly sensitive aspirating smoke detector, two gas sensors (H_2 and C_XH_X) and the microwave radiometer are detailed and detection algorithms are described. A general overview about the project and the carrier platforms is presented.

Keywords: fire detection, gas sensors, microwave radiometer, smoke detector, UAV, remote sensing, fire scout, fireguard.

1 General outline

Fast and effective detection is a key factor in forest fire fighting. To avoid uncontrollable wide spreading of forest fires it is necessary to detect fires in an early state and to prevent the propagation. It is important to move adequate fire equipment and qualified operational manpower as fast as possible to the source of the fire. Furthermore, an adequate logistical infrastructure for sufficient supply with extinguishing devices and maintenance is necessary, as well as continuous monitoring of fire spread. Moreover, the training of personnel is an important component for successful combating of forest fires. An integrated approach for forest fire detection and suppression is based on a combination of different detection systems depending on wildfire risk, the size of the area and human presence, consisting of all necessary parts, such as early detection, remote sensing techniques, logistics, training by simulation, and fire-fighting vehicles, see fig. 1.

Different risk levels, the size of the area and human presence define the applied sensing techniques. Small high risk areas can be observed by local staff. For very large and low risk areas satellite and aero monitoring is possible.

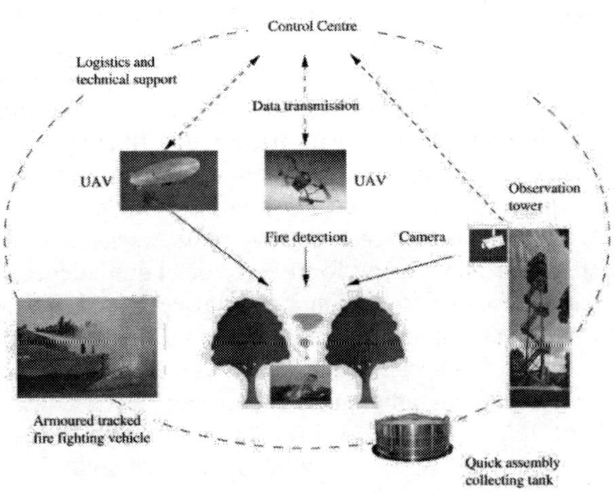

Figure 1: Schematic structure of the integrated forest fire detection and fire fighting system.

Figure 2: Scenes with alarm and false alarm situations.

Especially in the eastern part of Germany, several hundred observation towers equipped with camera-based systems have been built to observe forests. Recorded video sequences are transmitted to a control centre and analyzed by appropriate software. If a fire is clearly identified, fire suppression will be initialized by an alarm going directly to the fire brigade.

As known from other fire detection technologies the problem of false alarms requires additional measures for alarm verification. Possible false alarms are caused e.g. by dust produced by farmers, pollen, fog, smoke and water plumes produced by power plants, fig. 2. In case of an ambiguous decision a remote controlled **U**nmanned **A**erial **V**ehicle (UAV) equipped with gas sensors and an infrared camera can fly to the place where a fire is assumed to provide detailed pictures and a multitude of other measured data. Depending on these data a fire will be confirmed or unconfirmed.

If the source of a fire is not accessible for common fire fighting vehicles with conventional tyres it is only possible to extinguish the fire from the air or by so-called "smoke jumpers". A new solution is a reconstructed armoured and tracked fire fighting vehicle as known from military use, supporting existing airborne fire fighting ground-based right at the source of the fire in tough, rugged terrain. An innovative extinguishing system with high-pressure vortex technology was developed to provide a vehicle with extremely low water consumption [1].

2 Carrier platforms for the detection system

A very high reliability of fire detection and a concomitant low false alarm rate can be achieved by the combination of an infrared camera, a microwave radiometer and additional sensors of fire products respectively smoke particles, working with conventional and established technologies of fire detection. These detectors should be insensitive to dust and water particles. Depending on the application a selection of these sensors will be mounted on a mini-drone or a blimp. For both platforms, available space and weight are limited.

2.1 Early forest fire verification with a mini-drone

Carrier platform for the detection system in case of early fire detection is an autonomously flying video drone with excellent navigation skills in every territory. The AirRobot AR100-B (fig. 3a) offers a budget priced alternative to the normal aerial surveillance by a manned helicopter. The UAV enables the fire brigade to have maximum awareness of the situation and the occurrences during their mission. It can either be used for confirmation of an alarm detected by a video system or as well as a scout, helping to find hot spots especially at night while no fire fighting planes are flying. The UAV has a diameter of one meter, a weight of about one kilogram and a silent electric drive (4 brushless- and gearless DC-motors). It can be flown without any pilot experience. Data transmission and control in real-time is possible by RF devices. The telemetry-data is shown live on the ground station and can be tracked on a map in real-time. The whole processing for obstacle detection and collision avoidance is taking place autonomously in the AirRobot.

2.2 Hot spot surveillance with a blimp

In the case of monitoring extinguished fires, observation with a radiometer, gas and smoke detectors and a thermal camera attached to a blimp (fig.3b) is

(a)

(b)

Figure 3: (a) AirRobot UAV. (b) Blimp as a fireguard.

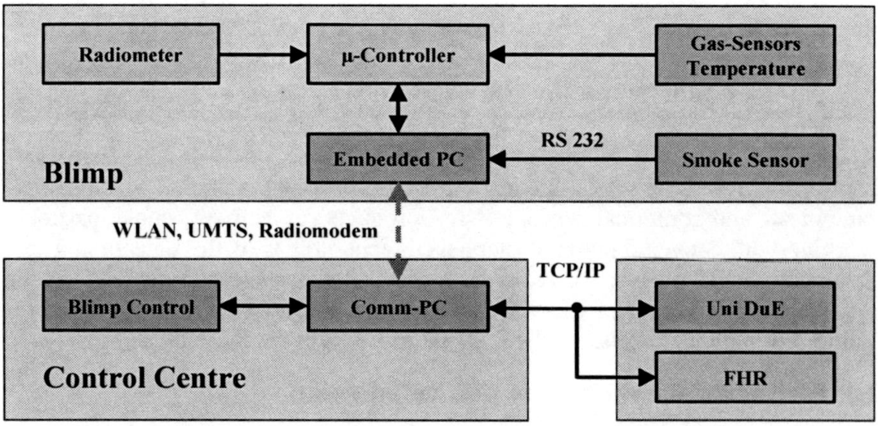

Figure 4: Communication structure for the blimp.

performed. Benefit of the 9 meter long airship with 2.3 m diameter is a high payload of 7 kg including batteries. The embedded-PC shown in fig. 4 controls the blimp as well as the communication and data transfer to the ground station.

The ground station consists of a user interface (Comm-PC) to transmit and receive flight information from the blimp via WLAN. Users (Uni-DuE, FHR) have to request their sensor data via TCP/IP at the ground station. The received data are combined with GPS data of the blimp and a time stamp for visualization and fed into the detection algorithms, determining whether an alarm is executed or not.

3 Early fire detection with gas sensors and smoke detectors

The sensor system shall be used to verify an ambiguous situation detected by a video-based system as well as observing an extinguished fire. Therefore sensors have to be widely immune against disturbances like steam, fog, dust pollution and condensing water, which cause video-based systems to give false alarms.

If fire gases are carried to the detector by the airflow, they are analysed with different semiconductor gas sensors. A gas-permeable protective cap made of sintered metal protects the sensor elements against soiling with dust and humidity. Conditional on its principle the sensor array is not affected by nuisance aerosols like dust, dirt, mist or condensing water [3].

Early forest fire detection sensors have to fulfil a lot of specific requirements compared to conventional applications. High sensitivity is needed to detect even low smoke concentration; dilution and extreme turbulence caused by wind are essential factors. Due to high occurrence of hydrogen during an open fire a H_2-Sensor [0 - 10ppm] was implemented [4, 5]. Main features of this semiconductor gas sensor (GTE GSME) are a very fast response time and a high sensitivity [3]. A C_XH_X-Sensor [0 - 5ppm] is used because hydrocarbon sensors are sensitive to organic fire products. Fast temperature fluctuations are measured by a temperature sensor. For smoke detection a high sensitive aspirating system (Hekatron ASD535 [6]) is used to detect even low smoke concentrations. The structure of the implemented detection system is shown in fig. 5.

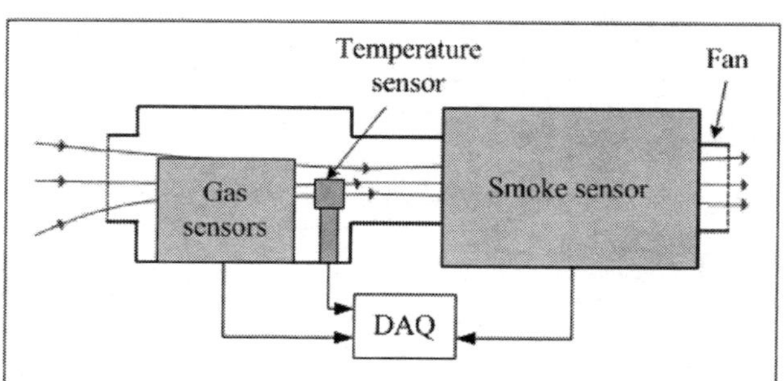

Figure 5: Structure of the sensor system and the internal airflow.

4 Detection algorithm for gas and smoke detection

With the analysis of sensor data measured by the AirRobot drone a pre-alarm of a video-based system shall be confirmed or unconfirmed. The measured sensor signals for hydrogen (N_{H2}), hydrocarbon (N_{CH}) concentration and temperature (N_T) are transmitted to the control centre and fed into the detection algorithm, according to eqn. (1, 2). The decision "alarm / no alarm" is the result of the output value $S_{gas,T}$ compared to the alarm threshold. Due to the application and the limited payload of the UAV a smoke detector is not implemented.

The observation of extinguished areas is performed by a blimp in order to detect buried hotspots and reignited seats of fire. This application requires an additional smoke detector and the detection algorithm has to be enlarged, according to eqn. (3, 4). The measured sensor signals for smoke density (N_D), hydrogen (N_{H2}), hydrocarbon (N_{CH}) and temperature (N_T) are transmitted from the blimp to the control centre. Fig. 6 shows the structure of the implemented detection algorithm.

$$S_{gas,T} = [(N_{H2,filtered}\, K_{H2} + N_{CH,filtered}\, K_{CH})\,(1 + N_{T,filtered}\, K_T)] \tag{1}$$

$$S_{gas,T} \geq \text{Alarm threshold} \Rightarrow \text{Alarm} \tag{2}$$

The algorithm can be adapted to the environmental conditions by subtraction of the sensors' quiescent values; just variations will cause an alarm. Pulse forming is necessary because of different response characteristics of the sensors. Especially in situations with high turbulences with very short smoke and gas pulses it is helpful. The third part of the algorithm is weighting and fusion of the pre-processed sensor signals.

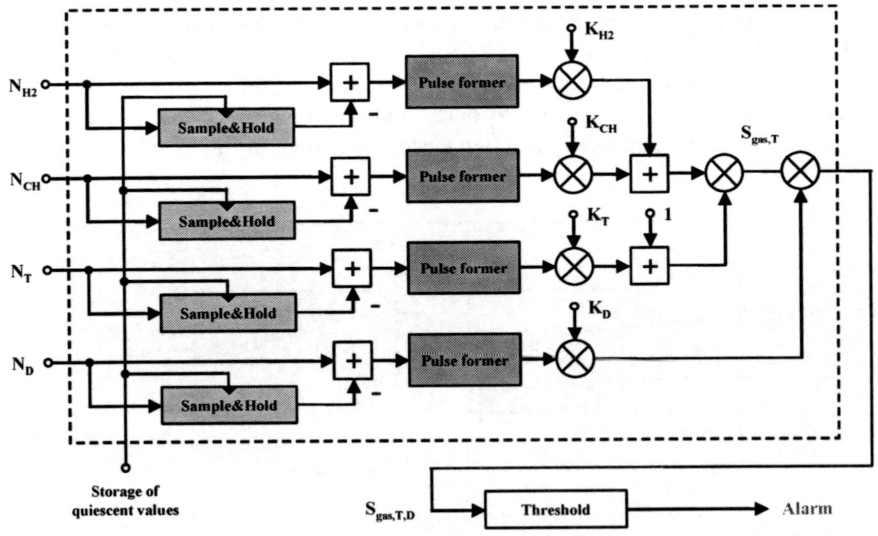

Figure 6: Structure of the algorithm.

$$S_{gas,T,D} = S_{gas,T}\,(N_{D,filtered}\,K_D) \quad (3)$$

$$S_{gas,T,D} \geq \text{Alarm threshold} \Rightarrow \text{Alarm} \quad (4)$$

The gas production of open and smouldering fires is different. Due to the chosen weight (K_{H2} und K_{CH}) of the gas sensor signals the susceptibility of the algorithm to both types of fires is similar. The algorithm also includes the temperature and the light scattering measurement.

5 Microwave radiometer

The proposed microwave radiometer detects fire radiation at 22.3 GHz. According to Planck's law a black body reaches its maximum of the radiation power in the infrared region. With the assumption that fire has similar characteristics as a black body, the wide use of conventional IR-cameras in fire detection applications is justified. In [7] it was shown that fire detection is also possible with microwave sensors. Additionally, good transmission behaviour in smoke, dust and fog is characteristic for sensors working in the lower GHz region [8]. There are long wavelengths in the background, which lead to less absorption and scattering in particles of the materials compared to infrared light. This characteristic of microwaves can be exploited for hot spot detection in smoky environments. Furthermore, after extinction of a forest fire, hot spots often occur and propagate under the upper surface of the earth, which often causes reignition of forest fires. A microwave radiometer allows detection of hot spots even if covered by thin layers of leaves, scrub and the like.

Fig. 7 presents the principle of the radiometer. Via a switch a signal is alternately received from the scene of interest (through a planar group antenna) and a noise source. This can be exploited for signal normalisation. In the

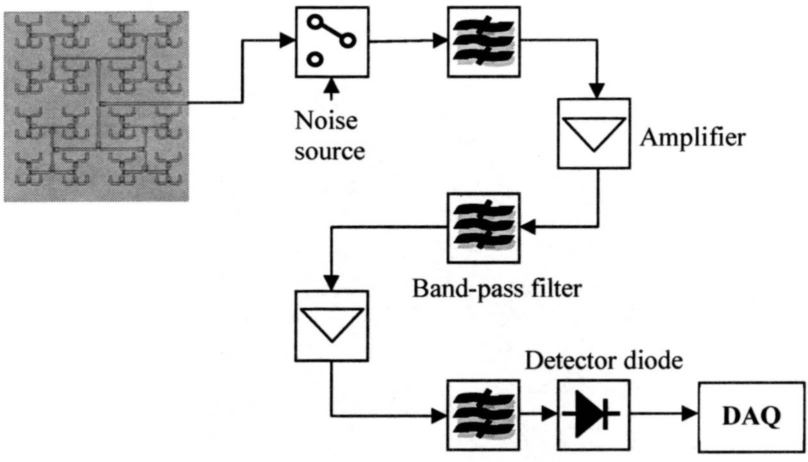

Figure 7: Block diagram of the microwave radiometer.

following processing the signal gets amplified and filtered. To prevent disturbances of the radiometric signal by other systems, the frequency of the radiometer was adapted to a radio-astronomical band with a bandwidth of 600 MHz. For data processing in the DAQ the signal must be converted from power into voltage using a detector diode.

The image resolution of the radiometer is dependent on the antenna opening angle and the flying altitude and follows basic trigonometric rules. The antenna opening angle can be calculated from frequency f and antenna size d, eqn. (5), c_0 being the speed of light in a vacuum.

$$\alpha = 72 * \frac{c_0/f}{d} \tag{5}$$

For the frequency of 22.3 GHz, a flying altitude of 30 m, an antenna size of 20 cm and an opening angle of approx. 5°, radiation can be detected from a square cell of 2.6 m edge length, the so called footprint of the system. The detection probability of a fire is highest if the footprint overlaps the fire completely. With the radiometer described above, only one pixel is generated. To produce several pixels, more than one antenna, a scanning system or a systematic flying platform must be implemented.

Fig. 8 shows the radiometer together with a temperature sensor, processing boards (PCB) and DAQ. The size is 105 mm x 150 mm x 73 mm and the weight including the case (300 g) is approximately 800 g. Further reductions are possible by designing chip based high frequency components and merging all single boards.

To increase the field of view of the radiometer the planar group antenna is attached to a small video camera with vertical and horizontal scan axis, fig. 9. Both sensors get data from the same direction. These data sets can be superimposed.

Figure 8: Image of the 22.3 GHz radiometer without antenna.

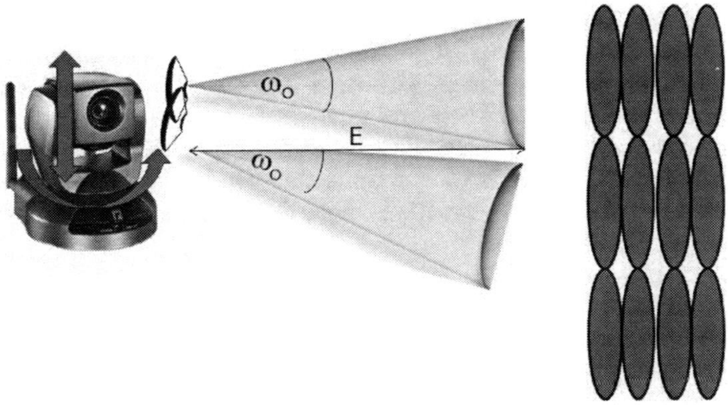

Figure 9: Scanning mechanism and illustration of footprints on the ground in distance E.

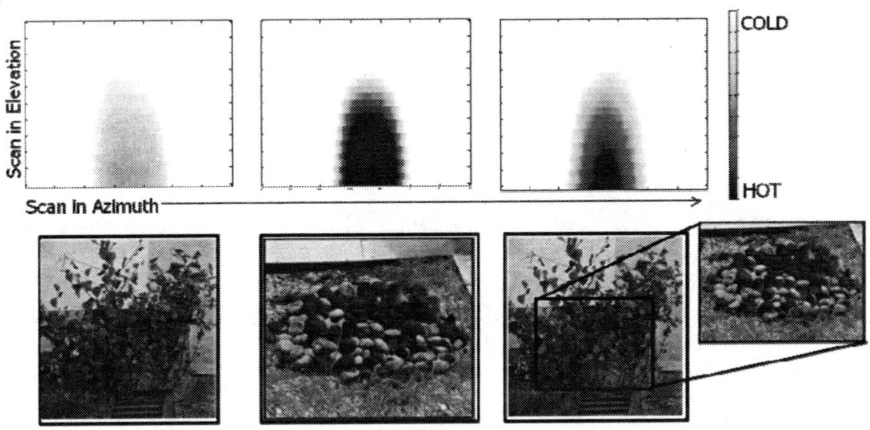

Figure 10: Radiometric measurement; left: tree alone, centre: charcoal alone, right: tree in front of charcoal fire.

The results of a 2D scanning radiometer and the good transmission characteristics of microwave radiation are demonstrated in fig. 10 by means of the transmission through a small tree. The radiometer was placed on a turntable to receive 2D images. The image to the left shows a scan of the tree alone, the image in the middle the radiation of a charcoal fire. On the right-hand side, the tree was placed in front of the fire. It can be seen that the radiation of the fire is attenuated in the tree but that it can still be clearly identified.

6 Outlook

A good interrelation of a very early and reliable smoke detection of forest fires, remote sensing techniques, logistics and technical support, training of fire fighters by simulation and an adequate extinguishing and rescue system will reduce damage and smoke impact on humans. The main focus has to be on early smoke detection because large and high-intensity forest fires are widely uncontrollable and cause very high risks.

To reduce false alarms of video-based systems, especially in hardly accessible terrain, a remote controlled UAV can fly to the place where a fire is assumed to confirm that the origin of the smoke is most likely a fire. After extinguishing a blimp can work as a fireguard.

The 22.3 GHz radiometer is able to detect fire even under insufficient vision due to smoke emissions, dust or fog. Materials as leaves and thin walls can be partly transmitted by the microwave radiation. The current design can be further miniaturized using chip-based components.

Semiconductor gas sensors are widely immune against disturbances like steam, fog, dust pollution and condensing water. Together with an aspirating smoke detection system it is possible to detect even lowest smoke and gas concentrations to confirm an alarm under laboratory conditions as well as under outdoor conditions. Finally gas concentration, smoke density, results of radiometric measurement and pictures of a thermal imaging camera enable a good discrimination between alarm and false alarm.

References

[1] Henrichs, M., *Armored and Tracked Vehicle for Rescue / Extinguish / Defend Missions*, 14th International Conference on Automatic Fire Detection, AUBE '09, Duisburg, Germany, September 8-10, 2009.

[2] von Wahl, N., Heinen, S., Tobera, R., Nüßler, D., Brauns, R., Schröder, M., Knott, P., Krüll, W. & Willms, I., *Intermediate Report Internationale Waldbrandbekämpfung iWBB, FHR-Report Nr. 134,* FGAN Research Institute for High Frequency Physics and Radar Techniques, Wachtberg, Germany, 2009.

[3] GTE, http://www.adicos.de

[4] Krüll, W., Willms, I., Tobera, R. & Wiggerich, B., *Early forest fire detection and suppression - an integrated approach*, 14th International Conference on Automatic Fire Detection, AUBE '09, Duisburg, Germany, September 8-10, 2009.

[5] Tobera, R., Krüll, W. & Willms, I., *Optical smoke and gas sensors as an additional method for early wildfire verification*, 14th International Conference on Automatic Fire Detection, AUBE '09, Duisburg, Germany, September 8-10, 2009.

[6] Hekatron, http://www.hekatron.de

[7] von Wahl, N., Heinen, S., *Advantages of millimeter waves in fire detection and monitoring,* 14th International Conference on Automatic Fire Detection, AUBE '09, Duisburg, Germany, September 8-10, 2009.

[8] Nüßler, D., Essen, H., von Wahl, N., Zimmermann, R. Rötzel, S., Willms, I., *Millimeter Wave Propagation through dust*, SPIE conference proceedings, September, 2008.

Assessing burn severity using satellite time series

S. Veraverbeke[1], S. Lhermitte[2], W. Verstraeten[3] & R. Goossens[1]
[1]*Department of Geography, Ghent University, Belgium*
[2]*Centro de Estudios Avanzados en Zonas Aridas, Universidad de la Serena, Chile*
[3]*Department of Biosystems, Katholieke Universiteit Leuven, Belgium*

Abstract

In this study a multi-temporal differenced Normalized Burn Ratio ($dNBR_{MT}$) is presented to assess burn severity of the 2007 Peloponnese (Greece) wildfires. Eight-day composites were created using the daily near infrared (NIR) and mid infrared (MIR) reflectance products of the Moderate Resolution Imaging Spectroradiometer (MODIS). Prior to the calculation of the $dNBR_{MT}$, a pixel-based control plot selection procedure was initiated for each burned pixel based on time series similarity of the pre-fire year 2006 to estimate the spatio-temporal NBR dynamics in the case that no fire event would have occurred. The $dNBR_{MT}$ is defined as the one-year post-fire integrated difference between the NBR values of the control and focal pixels. As such, the $dNBR_{MT}$ accounts for both the direct fire impact and vegetation responses. The $dNBR_{MT}$, based on coarse resolution imagery with a high temporal frequency, has the potential to become either a valuable complement to fine resolution Landsat dNBR mapping or an imperative option for assessing burn severity at a continental to global scale.

Keywords: differenced normalized burn ratio, fire severity, burn severity, MODIS, multi-temporal, vegetation regeneration, remote sensing.

1 Introduction

Wildfires play an integral role in the ecological functioning of many ecosystems, as they partially or completely remove the vegetation layer and affect post-fire vegetation composition. The assessment of post-fire effects is a major challenge to understanding the potential degradation processes after a fire and to comprehend an ecosystem's post-fire resilience. To name these post-fire effects, the terms fire severity and burn severity are often interchangeably used [1], describing the amount of damage the physical, chemical and biological changes or the degree of alteration that fire causes to an ecosystem. Some authors, however, suggest a clear distinction between both terms by considering the fire disturbance continuum [2], which addresses three different temporal fire effects phases: before, during and after the fire. In this context, fire severity quantifies the short-term fire effects in the immediate post-fire environment, whereas burn severity quantifies both the short- and long-term impact as it includes response processes (e.g. resprouting, delayed mortality) [3, 4].

In remote sensing studies burn severity is traditionally estimated using fine resolution imagery [5]. A popular approach, partly because of its conceptual simplicity, can be found in ratioing band reflectance data. Both the Normalized Difference Vegetation Index (NDVI) and the Normalized Burn Ratio (NBR) are frequently used in this context; however, the latter has become accepted as the standard spectral index for assessing fire impact. The NBR relates to vegetation moisture content by combining the near infrared (NIR) and mid infrared (MIR) spectral regions, in which large post-fire reflectance changes occur. However, when using mono-temporal post-fire imagery, unburned sparsely vegetated areas and burned areas are often confounded [4]. Therefore pre- and post-fire NBR images are generally bi-temporally differenced, resulting in the differenced NBR (dNBR), which permits a clear contrast between burned and unburned regions.

The dNBR method relies on Landsat imagery and is thus dependent of fine resolution image availability, which is limited to infrequent images over small areas. As a result, fine resolution burn severity studies have proven to be valuable for obtaining detailed information over specific fires; however, they fail to provide a multi-temporal overview of burn severity on a regional to global scale. Moreover, the lag, i.e. the time since the fire, and seasonal timing of the Landsat bi-temporal dNBR assessment influence the burned pixels' dNBR variability and the absolute magnitude of change [4, 6]. With regards to seasonal timing, an assessment during the green and productive stage of the vegetation contains a higher variability in post-fire effects than an assessment made in the dry season [4]. With respect to lag timing, Veraverbeke et al. [6] demonstrated that three weeks post-fire was the best time to estimate the fire impact with the dNBR, because then the fire-induced reflectance changes were maximal, which also resulted in more optimal pre/post-fire pixel displacements in the bi-spectral feature space. These temporal dissimilarities greatly limit the comparison between bi-temporal dNBR assessments of different fires, especially when a comparison between different ecoregions is required [5]. The use of multi-temporal data could possibly allow this comparison, as they permit the integration of the temporal dimension in a more robust way. To date very few studies have implemented coarse resolution time series to assess burn severity. The aim of this study therefore is the introduction of a multi-temporal dNBR ($dNBR_{MT}$) MODIS burn severity assessment approach based on minimum NIR

NBR composites. The method accounts for both the direct fire impact and vegetation regeneration [3].

2 Data

The study area is situated in the Peloponnese peninsula, in southern Greece (36°30'-38°30' N, 21°-23° E). After a severe drought period, several large wildfires of unknown cause struck the area in the summer of 2007. The fires consumed more than 150 000 ha of coniferous forest, broadleaved forest, shrub lands (maquis and phrygana communities) and olive groves.

Level 2 daily Terra MODIS surface reflectance (500m) tiles covering the study area (MOD09GA), including associated Quality Assurance (QA) layers, were acquired from the National Aeronautics and Space Administration (NASA) Warehouse Inventory Search Tool (WIST) (https://wist.echo.nasa.gov) for the period 01/01/2006–31/12/2008. The study area was clipped and the NIR (centered at 858 nm), MIR (centered at 2130 nm) and QA layers were extracted. Subsequently, the daily NIR, MIR and QA data were converted in eight-day composites using the minimum NIR criterion. After compositing a local second-order polynomial function, also known as an adaptive Savitzky-Golay filter [7], was applied to the time series as implemented in the TIMESAT software [8] to replace bad QA observations. Finally, the NBR index was calculated as:

$$NBR = \frac{NIR - MIR}{NIR + MIR} \tag{1}$$

3 Methodology

Before applying the $dNBR_{MT}$ (see section 3.2) approach it was necessary to create a burned area map, to extract per pixel fire dates and to select control pixels that estimate the NBR's spatio-temporal behavior for the case that there would not have been a fire occurrence (see section 3.1).

3.1 Preliminary processing

The burned area and fire date were extracted based on the characteristic persistency of the post-fire NBR drop, similar to the algorithm of Chuvieco et al. [9]. The accuracy of the burned area map was verified using a burned area map derived from Landsat Thematic Mapper (TM) imagery [10]. MODIS burned area statistics were extracted windows of 10 by 10 km. These statistics were regressed against their TM equivalents, in which the TM data served as independent variable and the MODIS data as dependent variable.

To minimize external and phenological variations a pixel based control plot selection method [11] was implemented. This control pixel selection is based on the similarity between the time series of the burned pixel and the time series of its surrounding unburned pixels for a pre-fire year. The selection is based on the time series similarity between pixels during the pre-fire year 2006 and the averaged Euclidian distance *D* was used as dissimilarity measure:

$$D = \frac{\sqrt{\sum_{t=1}^{N}(NBR_t^f - NBR_t^x)^2}}{N} \quad (2)$$

where NBR_t^f and NBR_t^x are the respective burned focal and unburned candidate control pixel time series, while N is the number of observations in the pre-fire year (N=46).

Application of the pixel based control plot selection method strongly depends on the size of the contextual neighborhood around the focal pixel, which determines the number of unburned control pixels x, and the number of selected control pixels c, since the combination of x and c will determine how well control pixels provide the temporal profile of the focal pixel as its best estimate in the case of no fire occurrence. For each burned focal pixel f a user-defined number of unburned candidate pixels x was selected. The candidate pixels are selected by increasing the spatial neighborhood window around f with one pixel on each side till the user-defined threshold is met. Subsequently, a user-defined number of pixels c that show the lowest temporal dissimilarity D with f is selected from the x candidate pixels. Then, the dissimilarity D was defined for the averaged time series of the c pixels as the existence of a beneficial averaging effect that removes random noise in the time series has been perceived in previous research [11]. The dissimilarity D is calculated for different settings of x and c to identify the most optimal control pixel selection conditions.

3.2 Multi-temporal dNBR

Burn severity incorporates both short-and long-term post-fire effects on the environment [3]. Consequently, burn severity is a combination of immediate fire impact and the ecosystem's ability to regenerate. Based on these characteristics, we propose a multi-temporal dNBR ($dNBR_{MT}$) that integrates the difference between the NBR values of a burned pixel and its corresponding control pixel over time. Doing so the $dNBR_{MT}$ is defined as:

$$dNBR_{MT} = \frac{\sum_{t=1}^{N}(NBR_t^f - NBR_t^c)}{N} \quad (3)$$

where NBR_t^f and NBR_t^c are the respective burned focal and unburned control pixel observations, while N is the number of post-fire observations included in the study (here N=46 for one year) and t=1 is the first post-fire observation. Figure 1 illustrates the principle of the $dNBR_{MT}$. Dividing by the number of post-fire observations N normalizes the $dNBR_{MT}$ data to the same range as bi-temporal dNBR assessments.

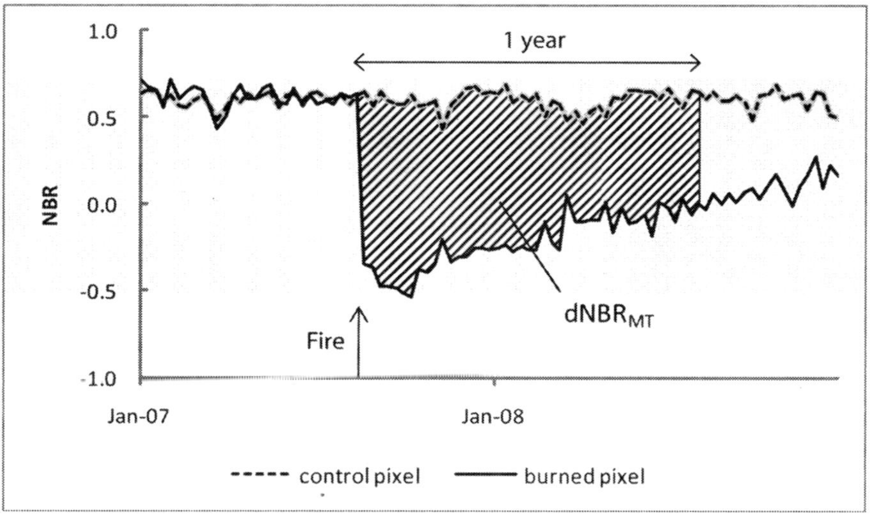

Figure 1: Principle of the multi-temporal dNBR ($dNBR_{MT}$). The $dNBR_{MT}$ represents the averaged integrated difference between the one-year post fire NBR time series of the control and focal pixels, as shown in the figure by the shaded area.

4 Results

4.1 Preliminary processing

Figure 2 displays the distribution plot and regression line of the TM-MODIS comparison. The MODIS–derived burned area map correlated fairly well with the Landsat-based map (coefficient of determination R^2=0.98, $p < 0.001$), although a consistent overestimation relative to the TM data was perceived as indicated by the regression slope of 1.31.

Figure 3 demonstrates the differences in mean dissimilarity D for different settings of the number of candidate and control pixels based on time series data of the pre-fire year 2006. Two main effects can be observed in the graph. Firstly, there was a consistently decreasing trend in dissimilarity D when the number of unburned candidate pixels increased. This feature appeared regardless of the number of control pixels chosen. Secondly, the number of control pixels chosen also influenced the dissimilarity measure due to an averaging effect. The strength of this averaging effect was again dependent of the number of candidate pixels: the more candidate pixels, the more important the averaging effect. The averaged time series of the five most similar control pixels resulted in a mean dissimilarity D of 0.0073. The results are satisfying since the mean absolute difference between the NBR values of the control pixels and the focal pixels was lower than 0.05 (eqn 2). We used this setting to calculate NBR images of the years 2007 and 2008, as best estimates of the spatio-temporal NBR dynamics in the case of no fire.

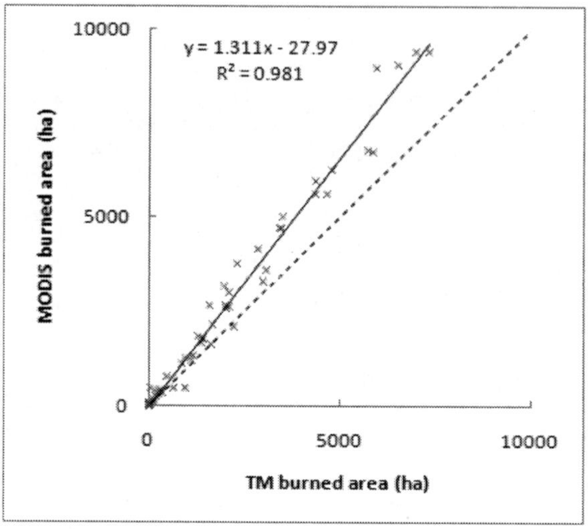

Figure 2: Distribution plot and regression line of the TM-MODIS burned area map comparison (regression significance of p<0.001).

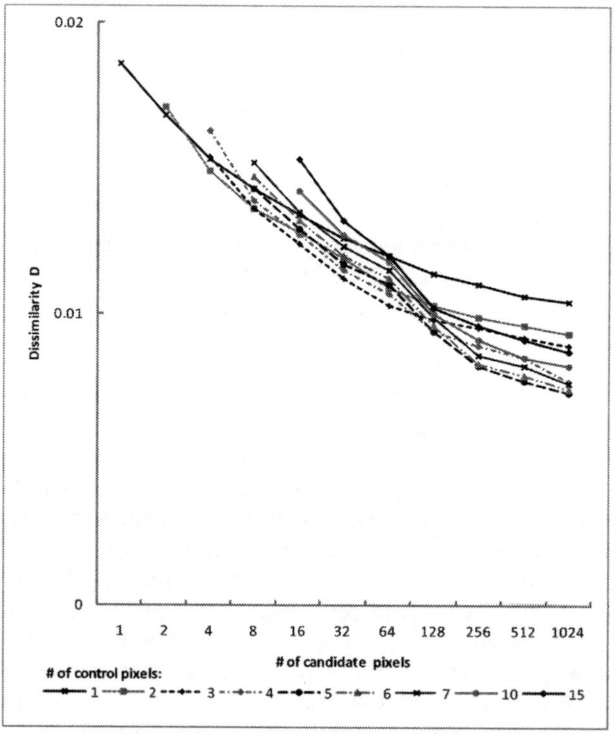

Figure 3: Mean dissimilarity D in function of the number of candidate and control pixels calculated based on NBR time series data of the pre-fire year 2006.

4.2 $dNBR_{MT}$

Figure 4A illustrates the result of the multi-temporal burn severity approach, while figure 5B details a specific burned area framed in blue in figure 4A. The images are mapped in a yellow-red color scale. Low $dNBR_{MT}$ values are represented in yellow, while a red color was assigned to high severity pixels. The spatial pattern of burn severity is clearly visible in the maps with an evident differentiation between high and low severity patches with relative smooth inter-patches transitions. The mean $dNBR_{MT}$ equaled 0.27 (sd = 0.14) (figure 4C).

Figure 5 shows the temporal profiles of a low, a moderate and a high severity pixel, A, B and C, respectively and their corresponding control pixels. These figures demonstrate that burn severity can be regarded as a combination of fire impact and the ability to recover. Although the three pixels experienced a clear NBR drop in the first post-fire weeks, the rate of recovery remarkably differed among the three pixels. Figure 5A reveals a very fast regeneration after the initial NBR drop. Half a year after the fire, NBR values of the burned pixel almost approached the control pixel's values. In contrast, in the example of figure 5C the burned pixel illustrated a more gentle post-fire increase in NBR. In this respect, the example pixel of figure 5B showed a somewhat intermediate behavior.

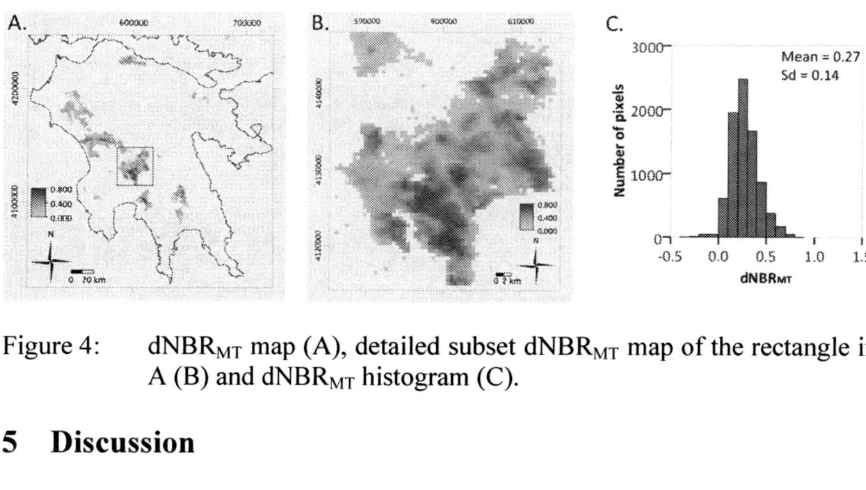

Figure 4: $dNBR_{MT}$ map (A), detailed subset $dNBR_{MT}$ map of the rectangle in A (B) and $dNBR_{MT}$ histogram (C).

5 Discussion

5.1 Preliminary processing

The regression model between TM and MODIS burned area statistics resulted in a high correlation (R^2=0.98) (see figure 3). Many authors have used fine resolution imagery to validate coarse resolution burned area maps (a.o. [12]). In some of these studies the coarse resolution approach produced an underestimation of the area burned relative to the fine resolution estimates (a.o. [12]), while in others an overestimation was perceived (a.o [12]). Although also related to vegetation type, the over- or underestimation can be explained by the burned area mapping algorithm's sensitivity to detect partially burned pixels

[12]. If partially burned pixels are easily detected, this will result in an overestimation of the total burned area. In our study a clear overestimation was observed as indicated by the regression slope of 1.31. This overestimation is due to the relatively relaxed threshold applied in the bi-temporal differencing of NBR images. This minimized the omission error, at the expense of a higher commission error. The control pixel selection procedure, however, requires a total exclusion of burned pixels, even when they are only partially burned. A more precise threshold would favor the selection of partially burned pixels as potential control pixels. This would result in a potentially flawed simulation of a control pixel's temporal NBR development in the post-fire stage [13].

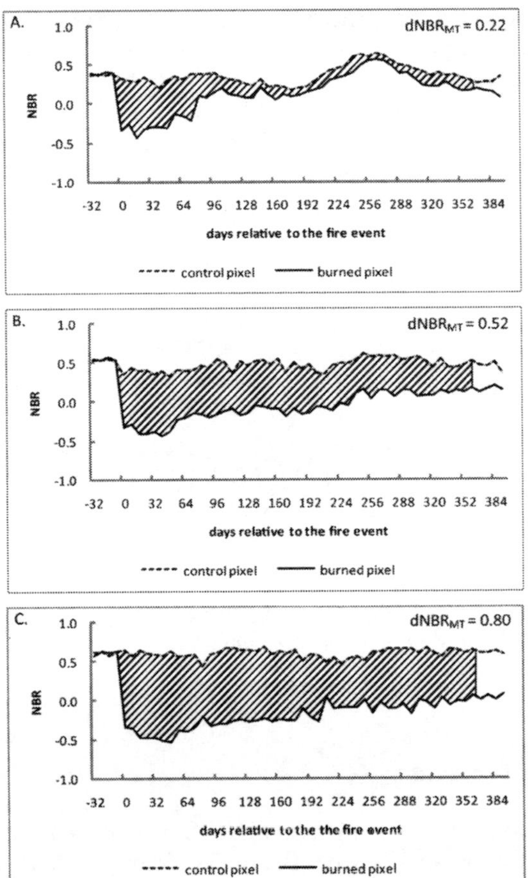

Figure 5: Illustration of low (A, 605600E 4131800N), moderate (B, 601400E, 4130400N) and high (C, 600900E, 4123500N) severity $dNBR_{MT}$ pixels. The area incorporated in the $dNBR_{MT}$ calculation is shaded.

The pixel-based control plot selection using a one year pre-fire time series revealed two main effects. Firstly, mean dissimilarity D decreased for an increasing number of candidate pixels, regardless of the number of control pixels chosen (see figure 3). This is trivial as by enlarging the spatial neighborhood around the focal burned pixel the probability of finding more similar pixels increases [11]. The relative decrease in dissimilarity is smaller when the number of candidate pixels doubles. This means that it is more likely that the most similar pixel is found close by the focal pixel than further away [11]. This corroborates with the recommendation of Li et al. [13] to select reference plots in the proximity of the burned areas. This is trivial as it follows a general rule that terrain features in a close vicinity are more likely to be similar than distant features. A second effect visible in figure 3 is the beneficial averaging effect, which became more obvious for a larger number of candidate pixels. This is rather remarkable as one would expect that the use of only the most similar pixel would give the best result. However, noise reduces the similarity between the most similar control and the focal pixel. Even after preprocessing, a certain amount of random noise remains present in the dataset. Averaging the two or more most similar pixels causes a more temporally stable signal because random noise is averaged out resulting in a higher temporal similarity with the focal pixel [11]. This beneficial averaging effect is, however, finite as at a certain point non-similar pixels will also be included in the averaging process, which will deteriorate the similarity. The number of pixels that must be included in the averaging process to reach the lowest dissimilarity also depends on the number of candidate pixels as a consequence of the higher probability to find more similar pixels among a larger amount of candidate pixels. So for a certain number of candidate pixels there exists a balance between favorable averaging and inclusion of dissimilar pixels. For our case the optimal balance was found by averaging the five most similar pixels out of 1024 control pixels.

5.2 $dNBR_{MT}$

A major advantage of the multi-temporal burn severity approach is its combination of both the immediate fire impact and vegetation regrowth. As such, it is more tightly connected to the definition of burn severity. Key [4] stated that burn severity encloses both first and second order fire effects. The most important first order effect is the fire's vegetation consumption, while vegetation regeneration and delayed mortality are substantial second order effects. In that respect, Lentile et al. [3] specified that burn severity relates to the amount of time necessary to return to pre-fire level. As such, they made a clear distinction between burn and fire severity, where the latter term is restricted to the immediate fire impact only. Bi-temporal assessments can only partly contribute to a thorough interpretation of burn severity as they fail to include post-fire vegetation responses. As seen in figure 5 the manifestation of regeneration processes can be diverse. Furthermore the degree of ecological information in bi-temporal studies is highly dependent on the timing of the assessment [4, 6]. Thus, where fine resolution Landsat studies allow revealing a high spatial detail, they miss the comprehension of the temporal dimension of burn severity.

Therefore, the higher temporal frequency of coarse resolution imagery can either be a vital complement to traditional Landsat dNBR mapping of specific fires or an imperative alternative for the assessment of burn severity at continental to global scales. Another strength of the $dNBR_{MT}$ is its insensitivity to inter-annual phenological differences, which frequently distort traditional pre/post-fire detection schemes [14]. This characteristic is due to the pixel-based control plot selection procedure [11]. Comparing the condition of a burned pixel with how the pixel would have behaved in the case of no fire, as estimated by the selected control plot within the same image, largely reduces external influences.

One of the prime concerns of scientists involved with burn severity mapping is the categorization of the fire-affected pixels in severity classes [4]. This duty is, however, to some part problematic due to the difficulty to determine absolute threshold that are ecologically meaningful across several ecoregions [15]. A first attempt to ameliorate the comparison of burn severity in different fires was made by Miller and Thode [15], who proposed a relative version of the dNBR (RdNBR). This index assesses the degree of environmental change caused by the fire relative to the pre-fire conditions. Therefore, more than the absolute index, the RdNBR hypothetically allows a better comparison among different land cover types, especially in heterogeneous landscapes. The improvement was made clear for fires in conifer dominated vegetation types in California, USA [16], although in other studies the correlation between the RdNBR and field estimates of severity was weaker than observed with the dNBR [17]. Miller and Thode [15] explicitly state that the implementation of a relative index will allow a more direct comparison of severity between fires across space and time. Whether the hypothetical advantage of the relative index to account for spatial heterogeneity has an intuitive appeal, the index does not handle temporal differences which may be present among different assessments. This is due to two main effects. Firstly, vegetation recovery decreases the vigor of the fire-induced change with time [6, 17], especially in quickly recovering ecotypes such as in the Mediterranean. Secondly, the seasonal timing determines the vegetation productivity and wetness of both the control and burned plots which influences the absolute magnitude of change and the degree of variability in the dNBR data [4, 6]. These two effects potentially hamper any comparison of two bi-temporal dNBR assessments and the issue has also a close linkage with the recent confusion in post-fire effects terminology (fire severity, burn severity, ecosystem response, etc.) [1]. Therefore caution is advised when using the dNBR to monitor and compare trends in burn severity either in time or across regions. The $dNBR_{MT}$, however, integrates the temporal variability into one value. If the period to which the integration is applied remains the same for different fires, the multi-temporal approach truly has the potential to allow a better comparison of burn severity either in time or space.

6 Conclusions

MODIS daily MIR and NIR reflectance products were used in a multi-temporal burn severity study of the 2007 Peloponnese (Greece) wildfires. Images were first composited in eight-day periods and subsequently missing values were replaced by means of the adaptive Savitzky-Golay filter. Prior to the calculation of the $dNBR_{MT}$ the burned area was mapped using the post-fire NBR drop characteristic. Then, for each burned pixel, a control plot was selected based on time series similarity of the pre-fire year 2006. Thanks to the control plot selection procedure the multi-temporal burn severity approach remained unaffected by phenological differences. After this preliminary processing, the $dNBR_{MT}$ was calculated as the one-year post-fire integrated difference between the NBR of the control and focal pixels, averaged by the total number of observations. By using a one-year post-fire time period, the $dNBR_{MT}$ incorporates both the immediate fire impact and the regeneration processes. These regeneration processes, combined with seasonal effects, make any bi-temporal analysis dependent on the timing of the assessment and as a consequence the inter-comparability of dNBR assessments conducted in different fires suffers from this. The $dNBR_{MT}$, on the other hand integrates the one-year temporal variability into one value. The $dNBR_{MT}$, which is based on coarse resolution imagery with a high temporal frequency, has the potential to be either a valuable complement to fine resolution Landsat dNBR assessments or an imperative option for burn severity mapping at a continental to global scale with an enhanced comparability of different fires across space and time.

References

[1] Keeley, J., Fire intensity, fire severity and burn severity: a brief review and suggested usage. *International Journal of Wildland Fire,* **18**, pp. 116-126, 2009.

[2] Jain, T., Pilliod, D., Graham, R., Tongue-tied. *Wildfire,* **4**, pp. 22–26, 2004.

[3] Lentile, L., Holden, Z., Smith, A., Falkowski, M., Hudak, A., Morgan, P., Lewis, S., Gessler, P., Benson, N., Remote sensing techniques to assess active fire characteristics and post-fire effects. *International Journal of Wildland Fire*, **15**, pp. 319-345, 2006.

[4] Key, C., Ecological and sampling constraints on defining landscape fire severity. Fire Ecology, **2**, pp. 34–59, 2006.

[5] French, N., Kasischke, E., Hall, R., Murphy, K., Verbyla, D., Hoy, E., Allen, J., Using Landsat data to assess fire and burn severity in the North American boreal forest region: an overview and summary of results. *International Journal of Wildland Fire*, **17**, pp. 443-462, 2008.

[6] Veraverbeke, S., Lhermitte, S., Verstraeten, W.W., Goossens, R., Assessment of the temporal sensitivity of the differenced Normalized Burn Ratio (dNBR) to estimate burn severity. *Remote Sensing of Environment*, in review, 2010.

[7] Savitzky, A., Golay, M., Smoothing and differentiation of data by simplified least squares procedures. *Analytical Chemistry,* **36**, pp. 1627-1639, 1964.

[8] Jonsson, P., Eklundh, L., TIMESAT-a program for analyzing time-series of satellite sensor data. *Computers and Geosciences,* **30**, pp. 833-845, 2004.

[9] Chuvieco, E., Englefield, P., Trishchenko, A., Luo, Y., Generation of long time series of burn area maps of the boreal forest from NOAA-AVHRR composite data. *Remote Sensing of Environment*, **112**, pp. 2381-2396, 2008.

[10] Veraverbeke, S., Verstraeten, W.W., Lhermitte, S., Goossens, R., Illumination effects on the differenced Normalized Burn Ratio's optimality for assessing fire severity. *International Journal of Applied Earth Observation and Geoinformation*, **12**, pp. 60-70, 2010.

[11] Lhermitte, S., Verbesselt, J., Verstraeten, W.W., Coppin, P., A pixel based regeneration index using time series similarity and spatial context. *Photogrammetric Engineering and Remote Sensing,* in press, 2010.

[12] Silva, J., Sa, A., Pereira, J., Comparison of burned area estimates derived from SPOT-VEGETATION and Landsat ETM+ data in Africa: influence of spatial pattern and vegetation type. Remote Sensing of Environment, **96**, pp. 188-201, 2005.

[13] Li, M., Qu, J., Hao, X., Detecting vegetation change with satellite remote sensing over 2007 Georgia wildfire regions. Journal of Applied Remote Sensing, **2**, pp. 021505, 2008

[14] Verbyla, D., Kasischke, E., Hoy, E., Seasonal and topographic effects on estimating fire severity from Landsat TM/ETM+ data. *International Journal of Wildland Fire*, **17**, pp. 527-534, 2008.

[15] Miller, J., Thode, A., Quantifying burn severity in a heterogenous landscape with a relative version of the delta Normalized Burn Ratio (dNBR). *Remote Sensing of Environment,* **109**, pp. 66–80, 2007.

[16] Miller, J., Knapp, E., Key, C., Skinner, C., Isbell, C., Creasy, R., Sherlock, J., Calibration and validation of the relative differenced Normalized Burn Ratio (RdNBR) to three measures of fire severity in the Sierra Nevada and Klamath Mountains, California, USA. *Remote Sensing of Environment,* **113**, pp. 645–656, 2009

[17] Allen, J., Sorbel, B., Assessing the differenced Normalized Burn Ratio's ability to map burn severity in the boreal forest and tundra ecosystems of Alaska's national parks. *International Journal of Wildland Fire*, **17**, pp. 463-475, 2008.

[18] Veraverbeke, S., Lhermitte, S., Verstraeten, W., Goossens, R., Evaluation of pre/post-fire differenced spectral indices for assessing burn severity in a Mediterranean environment with Landsat Thematic Mapper. *International Journal of Remote Sensing*, in press.

Real time fire front monitoring through smoke with bi-spectral infrared imaging

J. M. Aranda, J. Meléndez, L. Chávarri & F. López
LIR – UC3M: Laboratorio de Sensores Teledetección e Imagen IR
Departamento de Física, Universidad Carlos III de Madrid, Spain

Abstract

Infrared (IR) imaging is a standard technique in forest fire detection. In previous works we have shown that it can be used to classify the fire scene into regions (of embers, flame, fire front, etc). However, this requires multi-spectral images and a complex post-processing. In this paper we show that a less precise but still powerful classification of fire scenes, for distances in the range of hundreds of meters, can be done with a much simpler procedure. A fire index is obtained from bi-spectral images in the medium IR with an extremely simple processing that can be performed in real time. This makes it possible to aid the decision makers in forest fighting, by locating the fire front and the places of fire re-ignitions, and by indicating flame heights. Images of a prescribed forest fire obtained in a field campaign have been analyzed to define and validate experimentally the fire index. Comparison with the results of classification by post-processing of multi-spectral images shows a good degree of agreement, demonstrating the effectiveness of the fire-index approach.
Keywords: infrared, bi-spectral, forest fires, monitoring.

1 Introduction

Infrared (IR) sensors are already commonplace in forest fire related applications [1]. Satellites have become standard for fire danger estimation, mapping of burned areas and follow-up of after-fire recovery, and ground-based platforms are used in many places to provide early fire detection. However, the potentialities of IR for fire monitoring and characterization have not been fully

WIT Transactions on Ecology and the Environment, Vol 137, © 2010 WIT Press
www.witpress.com, ISSN 1743-3541 (on-line)
doi:10.2495/FIVA100111

realized yet. The very intensity of fire IR emission makes it difficult sometimes to obtain unsaturated images, and most IR cameras are used simply to "see the fire" without a radiometric calibration that would greatly enhance the information they provide.

Moreover, when trying to obtain quantitative measurements, even a calibrated camera faces non-trivial difficulties, which stem from the very nature of forest fires: complex targets that change with time, with several different regions (flames, fire front, embers...) whose spectral emission profiles may be very different from that of a blackbody. Fires, in addition, are usually observed from long distances, and each region, with a different spectral profile, is affected by atmospheric absorption in a different way. Therefore, measured radiances may be quite different from those emitted, and, in particular, apparent temperatures may differ strongly from the real ones, even for a perfectly calibrated camera.

In previous works, a scene analysis was performed using multi-spectral images that made possible to distinguish between fire regions ("classes") [2]. This classification allows to take into account the different spectral emission profiles of the regions and consequently to take full advantage from camera calibration. To obtain physical magnitudes, like embers temperature or radiated power, becomes thus possible [3].

However, this scene analysis is a relatively complex process, which can only be made with a post-processing of the IR sequences of images acquired during the burn. It is very useful for fire studies, but it does not improve fire monitoring. What is needed to help forest fire fighting is a tool that can be operated in real time. To this end, only a schematic scene analysis is necessary, in order to distinguish hot embers from active fire areas (fire front and re-ignition places) and, if possibly, to give an indication of flame height.

In this work we show that this can be done with a medium IR bi-spectral system and an extremely simple processing, thus making it possible to operate in real time. The method is applied to a field test burn of a suppression fire, and is validated by comparison with the results of the full post-processing method applied to a multi-spectral 3-band system.

2 Classification of IR multi-spectral forest fire images

The problem of classifying a fire scene, i.e., of assigning each pixel to a meaningful "class" previously determined (for instance, fire front, embers, flames, or background) can be solved along the lines well established in the field of satellite remote sensing, where multi-spectral images are used to make "thematic maps" of land use, assigning each pixel to a specific class (for instance, water, sand, urban areas or specific crops).

In the standard form used in satellite remote sensing, this process involves an algorithm called "classifier". Pixels in an image that are known to correspond to specific regions in the "ground truth" are designated as pure representatives of that class (usually called "endmembers"). For an n-band multispectral system, a pixel can be visualized as a point in an n-dimensional "multispectral space". An image translates into an n-dimensional scatterplot: a cloud of points, one for each pixel, where pixels of similar spectral composition are neighbors. Points corresponding to the endmembers are fed to the algorithm in a process that is called "training the classifier" [4]. After training, the classifier runs over all the pixels of the image and assigns each of them to the class whose endmembers are more similar to it, according to some specific criteria. This process is tedious since it generally requires a lot of ground truth images to train the algorithm, in particular when the dimensionality of the data (number of bands) grows.

In previous works [2] we demonstrated that, for fires observed at short distances (a few meters), this method can be simplified by using only two bands in the medium infrared, one at the CO_2 emission band near 4.26 μm and other outside it. For these bi-spectral images, the multispectral space reduces to a 2-dimensional space, and the scatterplot is simply a 2D graph where each axis corresponds to the radiance in one band, and the spectral composition of each pixel can be appreciated visually. Pixels that clearly correspond to specific regions, like embers or flames, can be identified on the scatterplot and selected as the endmembers for that class. The classifier assigns then the rest of the pixels to a class, in our case, by a maximum likelihood criterion.

This method of "classification on the scatterplot" simplifies greatly the process, since no ground truth images are needed. This simplification is possible because of the large contrast in the spectral composition of "flame" and "embers" regions, due to the strong CO_2 emission band. However, as distance increases, the atmospheric CO_2 absorbs this band, and classification becomes more difficult. "Classification on the scatterplot" techniques applied to direct images don't work well for fires at distances of hundreds of meters [5]. The obvious way out of this difficulty would be to use more bands, but this means, in principle, to use the full process of classifier training. This would be a tiresome work with uncertain results, because fires are not standard targets, and a large array of multispectral images with well-defined endmembers is not available.

Fortunately, classification on the scatterplot can still be feasible by applying a technique called Principal Component Analysis (PCA) to the original multispectral data. This is a standard method of image processing [4] that produces uncorrelated bands by making linear combinations of the often highly correlated original multispectral bands. Each combination is a "principal component" and there are as many PC bands as original bands. The transformation is in fact a translation in multi-spectral space that takes the origin to the center of mass of the cloud of points, followed by a rotation in such a way to obtain a diagonal covariance matrix, i.e., in order to orientate the coordinate axes of the space along the main axes of the cloud. This means that the PC images are uncorrelated. In addition, PCs are ordered by decreasing eigenvalue of the covariance matrix, i.e., PC1 has the larger variance, PC2 the second largest variance, and so on. This means that most of the information is contained in the first PC bands.

Therefore, when working with multi-spectral images, it may be wise to perform a PCA and retain only the first PC bands. This strategy applied to three-band images of experimental field burns has made possible classification using the PC2 vs. PC1 scatterplot up to at least 550 m of distance [5].

Although this method is simple as compared to standard classification techniques, it still requires a lot of computational load, since PCs have to be calculated for each frame, and the endmembers have to be identified manually on the scatterplot. Therefore, it cannot be performed in real time. This, in practice, is not an important problem when the aim is to obtain physical parameters like radiated power, since this processing is inherently complex: in addition to classification, images must be geo-referenced and an area must be assigned to each pixel. However, real time operation would be a very convenient feature to aid in forest fire fighting and suppression. A feasible system for real time operation must use as few bands as possible and keep processing extremely simple. In particular:

- Standard classification techniques, which train the classifier with well-known points in the ground truth, are prohibitively complex.

- PCA should not be used, since the coefficients of the linear combinations making up the PCs must be recalculated at each frame.

- Simple "classification on the scatterplot" methods are not feasible, since they require to draw a scatterplot and to identify clusters for each frame.

The simplest approach is to use only two bands and define a fixed combination of them as a "Fire Index", in a similar spirit to the well-known Normalized Difference Vegetation Index. In the following sections we explore this approach, comparing the results of the full 3-band maximum likelihood classification with those of the simple Fire Index-based processing defined here.

3 Experimental measurements

Field measurements have been performed on a prescribed suppression fire on an 82 m x 135 m scrubland plot. The test was conducted and instrumented by CIFAL-LOURIZAN, within the framework of the Fire Paradox research project.

The fire was monitored from a distance of about 480 m with several imaging systems (figure 1): a bi-spectral system in the medium (MIR) and thermal (TIR) infrared regions; a high speed MIR camera that allows to combine several integration times to increase dynamic range without saturation, and a MIR multispectral system with a rotating four filter wheel. The fire was registered also with a standard video camera in the visible region as a reference. Figure 2 shows an example of simultaneous visible-MIR images.

In this work we study the IR images acquired with the multispectral MIR system. We only have used tree filters with wavelengths centered at: F2 = 3.7 μm, F3 = 4.0 μm and F4 = 4.7 μm, and a width at half maximum of about

Figure 1: Imaging systems used in the test fire. From left to right, high-speed MIR camera, MIR multispectral system, MIR camera (with visible video camera on top) and TIR camera. The last two cameras integrate a MIR-TIR bi-spectral system.

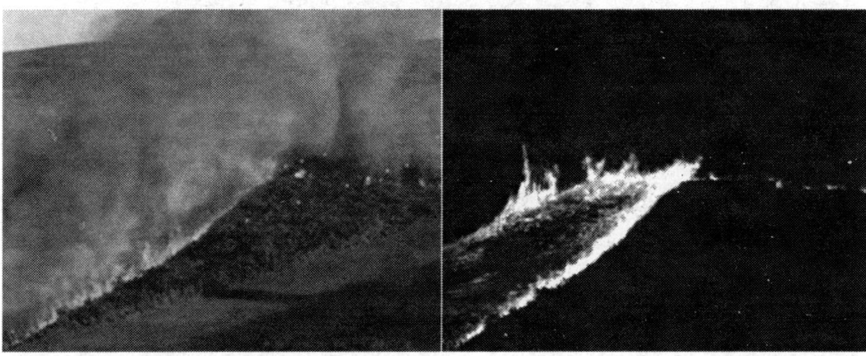

Figure 2: Two simultaneous images of the fire: visible (left) and 4.7 μm IR band (right).

400 nm. Position F1 was reserved to acquire in the entire MIR region, but it was not used to avoid saturation (the high-speed MIR camera images were used instead). The choice of these three filters is motivated by the spectral profile of CO_2, the main emitter of combustions in the MIR band. The strong emission band centered at 4.26 μm is divided by atmospheric absorption in two spikes: a narrow "blue" spike at short wavelengths and a wide "red" spike at long wavelengths. Filter F4 is centered at the longest wavelength region of the red spike, where atmospheric CO_2 absorption is presumably small; F3 filter is centered at the blue spike, and F2 is outside the CO_2 band.

4 Processing

A sequence of 20 images (one each 5 minutes) of the experimental burn described in the previous section was classified as explained in section 2. For

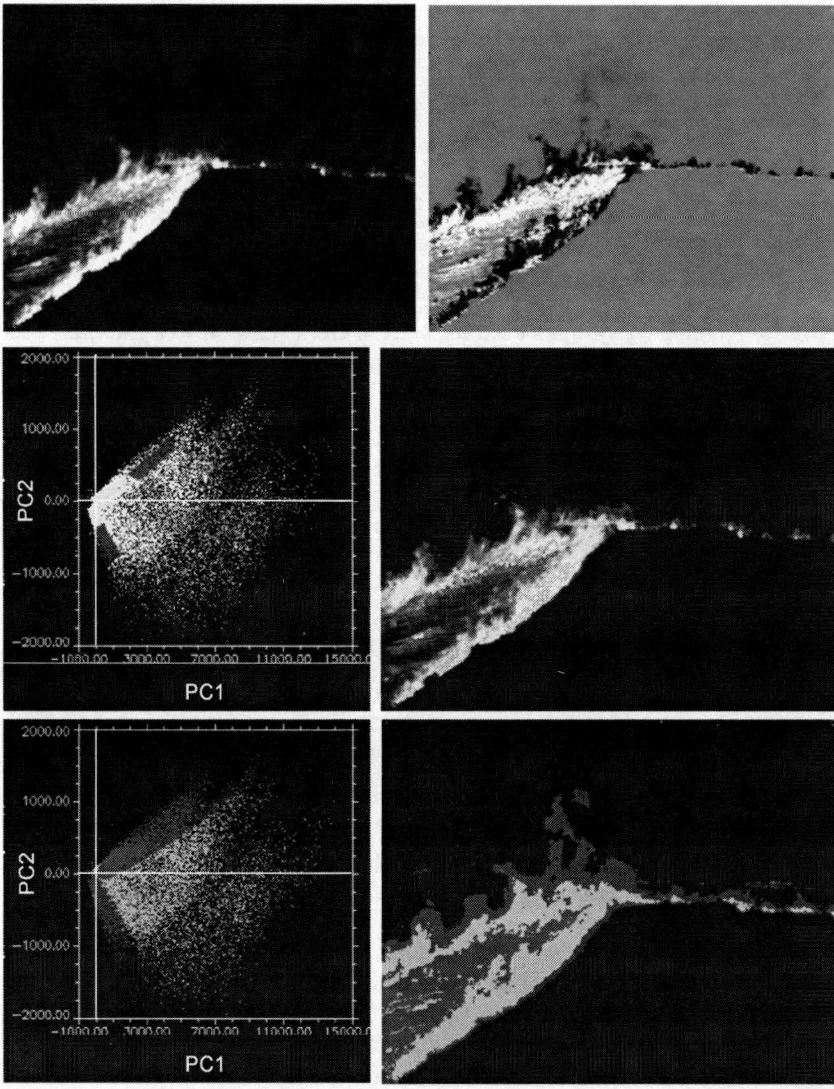

Figure 3: Top row: PC1 (left) and PC2 (right) for a typical frame during the burn. Middle row: PC2 versus PC1 scatterplot of the previous images with endmembers marked (left); position of endmembers on the image (right). Bottom row: Result of the classification on the scatterplot and on the image. The regions are: embers (green), fire front (yellow), flame (red) and background (blue) (color online only).

each frame, a PC analysis was performed; clusters for the endmembers of the different classes were identified on the PC2 vs. PC1 scatterplot and a maximum likelihood algorithm was used to assign the rest of the pixels to a specific class.

Figure 3 shows this process applied to a specific frame, number 3. In the top row, the first two principal components are shown. Their expression as combinations of the filters is:

$$PC1 = 0.698*f2 + 0.514*f3 + 0.498*f4$$
$$PC2 = 0.328*f2 + 0.389*f3 - 0.861*f4$$

where f2, f3, f4 stand for the digital number values of the F2, F3, F4 filters with their respective averages subtracted (so that over the image f2, f3 and f4 average to zero). Clearly, PC1 (shown at left) is a kind of mean of the three bands, measuring the overall IR brightness. PC2 (shown at right) adds the contributions of F2 and F3 but subtracts, with a larger weight, the contribution of F4. Since this filter collects the contribution of CO_2, flames appear with negative values (black in the image).

In the middle row of Figure 3, the PC2 vs. PC1 scatterplot is shown at left, with regions selected as endmembers: green for embers, red for flames, yellow for fire front (flaming embers) and blue for background. The right-hand image shows the location of the endmembers on the image. The result of a classification performed with those endmembers is shown on the bottom row of Figure 2. Each pixel has been assigned to the most similar class, in a maximum likelihood sense; at left it is shown the classified scatterplot; at right, the classified image. This process is repeated for each frame to obtain the classified images. We call this procedure Maximum Likelihood Three-Band (ML3B) classification.

Classification can be simplified if the linear combinations that give the PCs do not change very much as the fire evolves. This is indeed the case, as shown by figure 4 (top row), which shows the weights of PC1 and PC2 on the three filters along the burn. In addition, it is clear that filters F2 (3.7 μm) and F3 (4.0 μm) play a very similar role in PC1 and PC2. Differences in their value are only important for PC3, but this PC is not useful for classification.

Therefore, in order to obtain PC1 and PC2 filter F3 is redundant, and it seems plausible to use only filters F2 and F4 to study fires. The PC analysis has been repeated using only these two bands; the resulting PCs have the weights shown on figure 4 (bottom row). Again, weights are very similar along the whole burn.

This suggests a way to a very simple processing, using only two bands and two fixed combinations of them given by the nearly constant weights of the PC1 and PC2 calculated for those two bands. In this way we define two "pseudo-PCs" ("pseudo" because they are obtained by a fixed formula, not by an adaptive one as the real PCs):

$$psPC1 = 0.813*f2 + 0.582*f4$$
$$psPC2 = 0.582*f2 - 0.813*f4$$

To define the weights, we have calculated the PCs of a composite ("mosaic") image formed by all the frames; the average values subtracted from F2 and F4 to obtain f2 and f4 have been calculated also for this mosaic image.

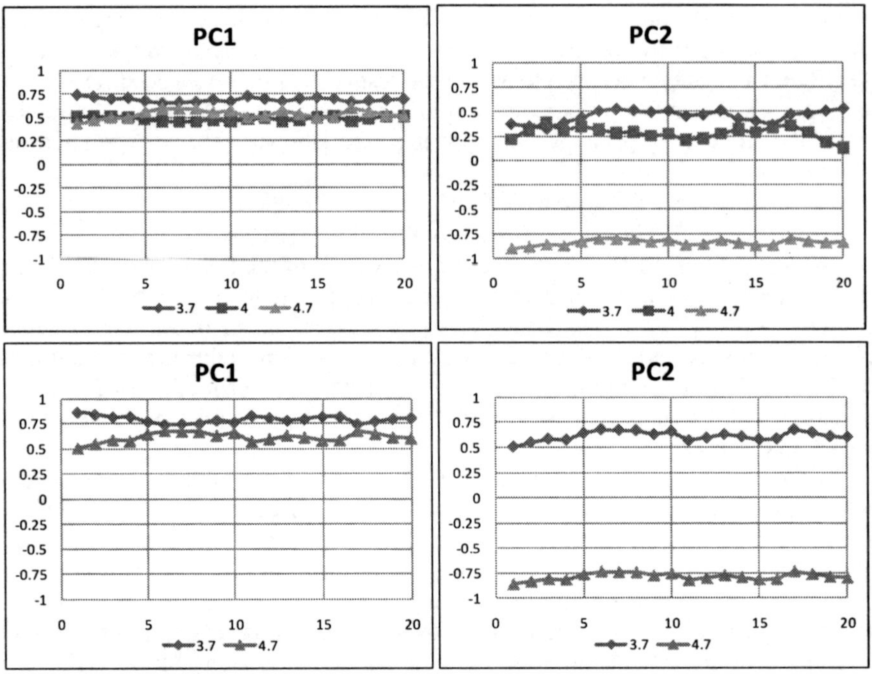

Figure 4: Weights of the two first PCs on the IR bands: Top, for three bands; bottom, for two bands.

These pseudo-PCs should be a good practical approximation of the real PCs for classification purposes. We have compared them graphically in figure 5. The left side shows the PC2 vs. PC1 scatterplot with the ML3B classes. At the right side, the same classes are shown on the psPC2 vs. psPC1 scatterplot. Although there are some differences on the shape of the scatterplot and the limits between classes are somewhat blurred, the overall pattern is very similar.

It is clear also that frontiers between classes in the scatterplot are quite straight. This suggests the idea of approximating them by straight lines that go through the origin and then to classify pixels according to the value of psPC2/psPC1. In fact, we will define the ratio psPC2/psPC1 as the Medium Infrared Fire Index:

$$\text{MIFI} = \text{psPC2}/\text{psPC1}$$

If the frontiers between embers and fire front and between fire front and flame are approximated, respectively, by lines with slope s1 and s2, then a pixel with psPC1 > 0 will be assigned to “embers” if MIFI > s1, to “fire front” if s1 > MIFI > s2, and to “flame” if s2 > MIFI. If psPC1 < 0, the pixel is considered as background.

A way to assess this fast “MIFI-classification” is by plotting the histograms of MIFI values for the ML3B classes over the whole fire studied. This has been done

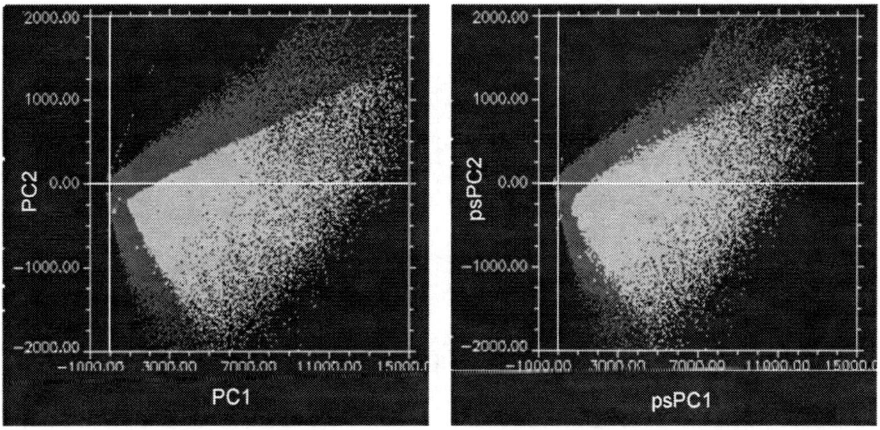

Figure 5: ML3B classes obtained for the whole burn plotted on the PC2 versus PC1 scatterplot (left) and on the psPC2 versus psPC1 scatterplot (right).

in figure 6 (left). Ideally, no overlapping peaks for each class should be obtained. It is clear that there is an important overlap between embers and fire front, but flame is well separated both from embers and fire front. This is in fact a great progress as compared to separability based on PC2 value only, as can be seen in figure 6 (right), where the fire front mixes completely with flame and embers.

Therefore, the MIFI provides a "fast classification" with acceptable results for practical purposes, and, since it applies a simple fixed linear combination to the digital numbers measured by the two F2 (3.7 μm) and F4 (4.7 μm) channels of a bi-spectral MIR imaging system, it can be implemented for real time operation in the field.

Figure 6 can be used to define the values of the threshold values s1, s2 used to classify the scene. For illustrative purposes we have used s1= 0.41, s2 = 0.0, although these values can be optimized depending on the relative importance of classification errors. An example of the results of this fast classification is provided by figure 7, where the ML3B classes (left) are compared with the MIFI classes (right) for frame number 5, showing a very good agreement.

Results can be summarized in the confusion matrix (table 1), which lists the percentage of pixels of each ML3B that is classified in each of the MIFI classes. Off-diagonal values are classification errors. Although there is an appreciable degree of crossing between the flame, embers and front classes, the overall accuracy is very good: a 96.5% of pixels correspond to the diagonal of the matrix; i.e., are classified equally by the two methods. If background pixels are excluded, accuracy is still 63.5%, which is an acceptable value taking into account that MIFI-based classification is extremely simple, and that s1, s2 threshold values are not optimized.

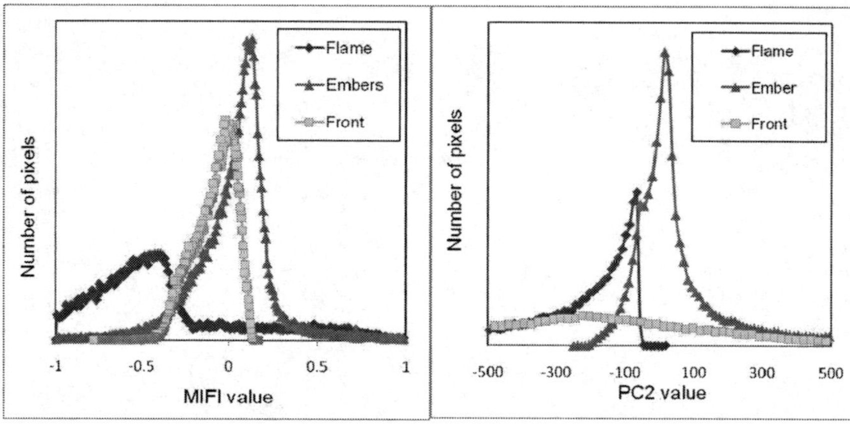

Figure 6: Histograms for the different ML3B classes of: MIFI values (left), PC2 values (right).

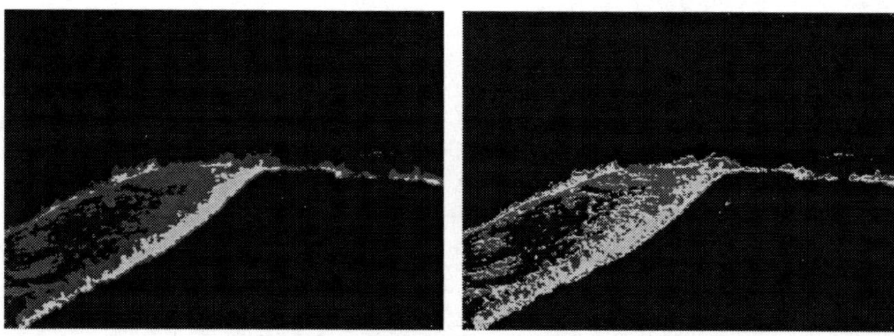

Figure 7: A typical image of the burn classified with ML3B classification (left) and classification based on MIFI values (right).

Table 1: Confusion matrix relating ML3B classification with MIFI classification.

	ML3B class (Percent)				
MIFI Class	Flame	Embers	Bkg	Front	**Total**
Flame	58.69	2.98	0.15	0.23	1.63
Embers	8.23	61.41	0.16	28.25	3.28
Bkg	10.82	3.35	99.49	0.02	91.49
Front	22.26	32.26	0.2	71.5	3.6
Total	100	100	100	100	100

5 Summary and conclusions

Studies of forest fires in the medium IR have demonstrated that the scene can be classified into different regions using the first two Principal Components (PCs) of multispectral three-band images, at 3.7, 4.0 and 4.7 μm. Analysis of the spectral weights of these components shows that they change little during the different fire stages, and that filter at 4.0 μm is redundant. Thus, pseudo-PCs have been calculated as fixed combinations of digital numbers measured at 3.7 and 4.7 μm, and the similarity with the real PCs has been assessed. Finally, a Medium Infrared Fire Index (MIFI) has been proposed that is the ratio of the two pseudo PCs: MIFI = psPC2/psPC1. This index is extremely simple in computational terms, so that it can be calculated in real time. The ability of MIFI to perform a fast classification at a distance of 480 m has been demonstrated by comparing classes obtained from threshold values in MIFI with classes obtained with a maximum likelihood algorithm that uses the real PCs calculated with the three original bands. A global accuracy of 96.5% has been found that indicates a very good agreement and demonstrates the effectiveness of the fire-index approach.

Acknowledgements

The authors wish to acknowledge J. A. Vega and E. Jiménez of CIF-Lourizan (Galicia) for the realization of the field tests. Finally, F. Díaz for assistance in the measurements and image processing. This work has been partially funded by Fire Paradox project FP6-018505.

References

[1] San-Miguel-Ayanz, J. & Ravail, N., Active Fire Detection for Fire Emergency Management: Potential and Limitations for the Operational Use of Remote Sensing. *Natural Hazards*, **35**, pp. 361-376, 2005.

[2] Aranda, J.M., Meléndez, J., de Castro, A.J., & López, F., Bi-spectral Infrared Forest Fire Detection and Analysis Using Classification Techniques *SPIE Proceedings*, **5153**, pp. 136-146, 2003.

[3] Meléndez, J., Aranda, J.M., de Castro, A.J., and López, F., Measurement of forest fire parameters with multi-spectral imaging in the medium infrared. *QIRT Journal*, **3**, pp. 183-201, 2006.

[4] J.A. Richards and X. Jia. *Remote Sensing Digital Image Analysis*, 3rd ed. Springer-Verlag, Berlin, 1999.

[5] Aranda, J. M., Meléndez, J., de Castro, A. J., and López, F., Measurement of physical parameter of forest fires by infrared imaging methods, *Proceedings of First International Conference on Modelling, Monitoring and Management of Forest Fire 2008*, pp. 111-120, Eds. J. de la Heras, C.a. Brebbia, D. Viegas & V. Leone. WTT Transactions on Ecology and the Environment, vol. 119 (2008).

Forestwatch® wildfire smoke detection system: lessons learned from its two-year operational trial

M. Lalkovič & J. Pajtíková
National Forest Centre, Slovak Republic

Abstract

In 2008 the wildfire smoke detection system Forestwatch® was introduced to Slovakia. The operational trial was set up at the State Forest Enterprise Kriváň and covered about 60,000 ha of mostly forested area. The trial benefited from the transfer of know-how from abroad. In 2008-2009, the trial focused on the system implementation and testing its feasibility in Slovak conditions. The trial results have fully confirmed its suitability for wider use in Slovak forests.
Keywords: fire monitoring, forests vulnerable to fire, wildfires.

1 Introduction

The Slovak republic is one of the most forested countries of Europe. Forests cover almost 41% of the country's area, which equals 1,934,000 ha. The proportion of coniferous and broadleaved species is 40.3% and 59.7%, respectively.

Every year fire damages extensive areas of forest stands. In Slovakia, fire prevention is currently performed by a combination of ground patrols of foresters and aerial monitoring, while the patrols are realised mainly during the weekends and holidays. Both methods are quite efficient, but rather limited in scope, particularly from the financial side. Due to the increasing topicality of this issue, it is important to search for and to apply alternative techniques of forest fire monitoring. The aim is to shorten the response time between fire occurrence and the start of its suppression while accounting for economic efficiency. One alternative possibility is to apply a fixed camera fire monitoring system.

WIT Transactions on Ecology and the Environment, Vol 137, © 2010 WIT Press
www.witpress.com, ISSN 1743-3541 (on-line)
doi:10.2495/FIVA100121

2 Fire detection system Forestwatch®

The fire detection system Forestwatch® is an automated fire detection monitoring system developed by the company EnviroVision Solutions PTY Ltd. from South Africa. In Europe, the product is distributed by Eagle Eye Protection from Greece, which is in Central Europe represented by ICZ Slovakia Ltd. This system is based on camera scanning of the area of interest and consequent evaluation of obtained images for the occurrence of smoke or fire signs. The system has over a 10-year history and has been applied in South Africa, Canada, USA, Chile, and other countries. The Forestwatch® system consists of several components fulfilling partial tasks to ensure the functioning of the system as a whole. The components are the following:

- Camera located at a tower (in our case we used model Pelco Esprit)
  - rotation mechanism (pan 360°)
  - tilt mechanism (tilt range from + 33° to -83°)
  - automatic focusing
  - 24x optical zoom
  - light sensitivity starting from 0.0005 lux
- Professional computer (ISE – Image Sampling Engine)
- communication subsystem
- Forestwatch® software

Figure 1: Control centre and camera.

The system has two main parts: towers with cameras, and a control centre. The Forestwatch® software is installed on the servers and computers in the control centre, while in the tower there is a camera, communication subsystem, and a professional computer (ISE). The system operates continuously 24 hours per day/7 days per week, when the camera scans the area of interest. The system is also able to perform night-time monitoring – the applied cameras have a high light sensitivity of 0.005 lux. Thanks to using 3D terrain models, the system can identify and position fires even in the areas without direct visibility, i.e. "behind the hill". Data collected by cameras are processed by the professional computer and transferred by the communication subsystem to the control centre for further processing. The software in the control centre receives, processes, and interprets the data. On the basis of the interpreted data, the operator of the control centre is alerted about the possibility of fire occurrence both visually and audibly.

The system has three alarm categories:

- new fire – the system alerts the operator to the fire in the area of interest
- old fire – the system signalises that in the monitored area there are still signs of an already identified fire
- unidentifiable condition – the system is not sure and needs the operator's interaction.

The operator has the possibility to take over the control of any camera and examine the indicated incident more thoroughly. The system works with a digital terrain model, in which the identified incidents are shown. Apart from the digital terrain model, the operator can also utilise a topographic map, data from cameras, and GPS coordinates of the incident. When the digital terrain model is used, no triangulation for the incident positioning is necessary.

The methodology of the operation is an integral component of the system. The methodology describes the work with the system, and consequent resolution of detected incidents, and utilises usual ways of fire reporting and communications with fire brigade and rescue service. The advantage of the system is the possibility to provide fire brigade with GPS coordinates of fire, which substantially facilitates and speeds up the process of fire positioning and suppression. The applied methodology can differ between the regions depending on local conditions and customs.

An example of resolving an incident can be described as follows:

1. fire occurs
2. system identifies the fire, a control centre operator is alerted visually and audibly
3. an operator takes over the control of the camera mounted on the particular tower
4. an operator performs a closer analysis of the situation, and identifies if the fire is under control or not
5. uncontrolled fire is reported to the control centre of Fire and Rescue Service

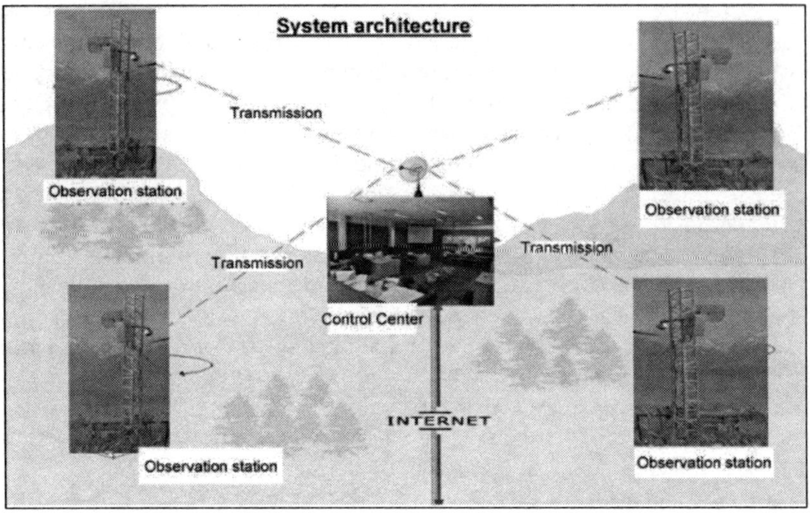

Figure 2: Scheme of functioning of system Forestwatch®.

3 Preparation, realisation and trial operation of the project

The project was realised with the support of the Ministry of Agriculture – forestry section, and on the base of the trilateral agreement between National Forest Centre Zvolen (coordinator), state enterprise Forest of the Slovak Republic (user), and ICZ Slovakia Ltd. (technology supplier). In addition, EnviroVision Solutions (Republic of South Africa) and Eagle Eye Protection (Greece) were cooperating in the project.

The operational trial was set up at the State Forest Enterprise Kriváň due to several reasons. The enterprise is situated in the area with higher fire risk, the majority of forests in the area are owned by the state, a great proportion of the area belongs to state nature reserve Poľana, and from the topography point of view the terrain is rather broken. Above all, the terrain ruggedness was an important feature for testing the system in Slovak conditions, since abroad the system has usually been applied in flat areas.

The system monitors the area of more than 60,000 hectares using 3 cameras. The cameras were positioned at mobile network towers. Due to the terrain configuration, it was needed to build one auxiliary tower at locality Ostrôžky to ensure data transmission. Within the framework of the trial operation, the system is also interconnected to the control centre of Fire and Rescue Service, which speeds up the response to fire occurrence. The connection is assured by multiple sources – telephone, e-mail, and web application.

In 2008, trial operation started on July 1st and lasted until October 1st 2008. During autumn and winter (i.e. from October 2008 to February 2009), the system was in so called sleep mode. It means that the system was in operation, but the

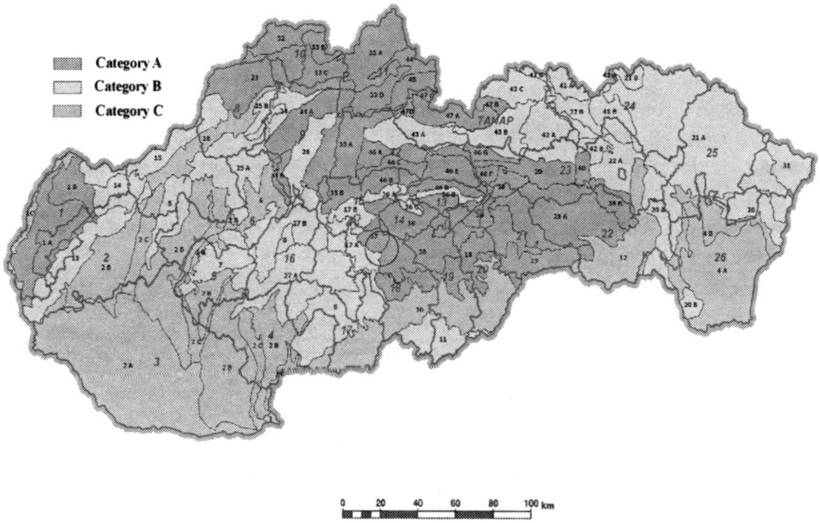

Figure 3: Regions of Slovakia according to the fire risk degree (dark grey/maximum, medium grey/minimum) and project location.

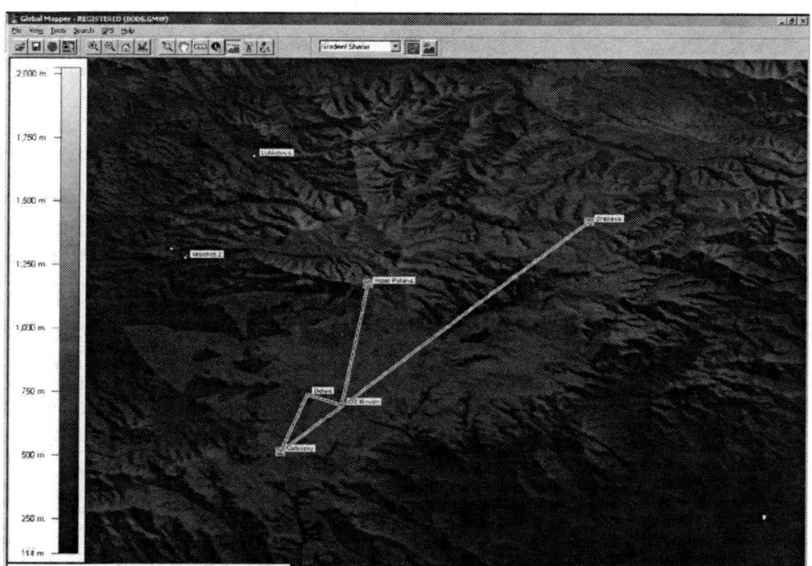

Figure 4: Preparation phase – visualisation of direct and indirect coverage of the area with cameras.

data were not processed because of very low or no fire risk. The new monitoring season was started in the middle of March 2009. Due to increasing fire risk, operators entered the system on May 15th 2009. The operation was terminated on October 31st 2009. Owing to financial problems, in 2009 monitoring was restricted to most critical periods of potential fire occurrence.

The successfulness and the suitability of the presented solution was determined on the base of the goals and the quantifiers specified in the project preparation phase and at its start as principal parameters, which could create a complex and a real picture about the possibilities and the risks of the utilisation of the given technology.

We accounted for

- The functionality of the system from the point of the needs of fire prevention
- The suitability of the system with regard to the area or locality
- The work methodology and service actions
- The system utilisation for other purposes apart from fire detection (e.g. illegal timber cutting, migration of wildlife, etc.)

Considering the personnel staff of the control centre, a so called mixed staff model was applied, i.e. during working days the operation of the control centre was realised by the employees of the Forest Enterprise, while during weekends and holidays by trained employees from forest districts. While the system was in operation, the staff was continually guided from the methodological aspect.

Figure 5: Poľana region – monitored area with marked mobile network tower, on which the camera is mounted.

4 Evaluation of the project realisation and operational trial

During the preparation and the realisation of the operational trial of the fire detection system no greater problems occurred, and the system was activated according to the planned schedule. The greatest requirements were put on the coordination of the activities, because, as it was already mentioned, apart from the three Slovak organisations two foreign partners from Greece and South Africa also participated in the project.

The actual fire detection process followed the applied methodology and fires could be evaluated in two basic categories:

- Controlled (reported) fires – identified by the system, planned, control centre (CC) was notified beforehand
- Uncontrolled fires – unplanned, identified by the system or an operator (grass burning, negligence, tourism...)

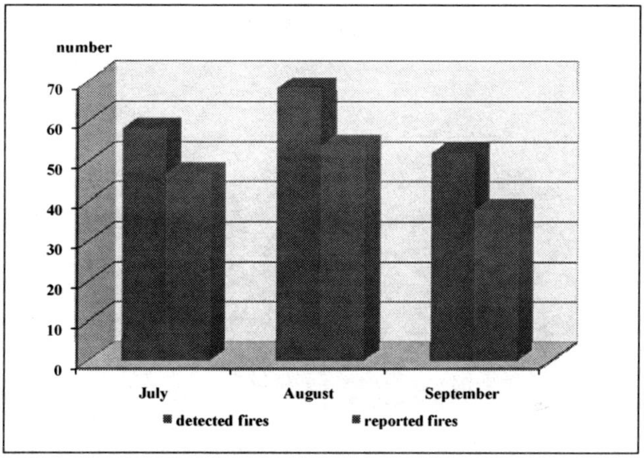

Figure 6: Fire occurrence in months of 2008.

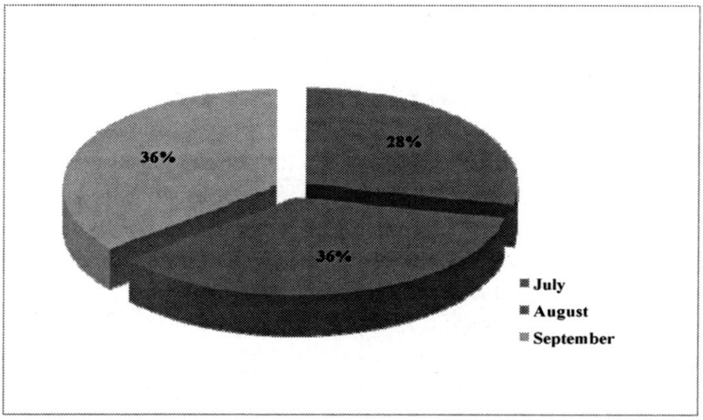

Figure 7: Uncontrolled fires in 2008 given in %.

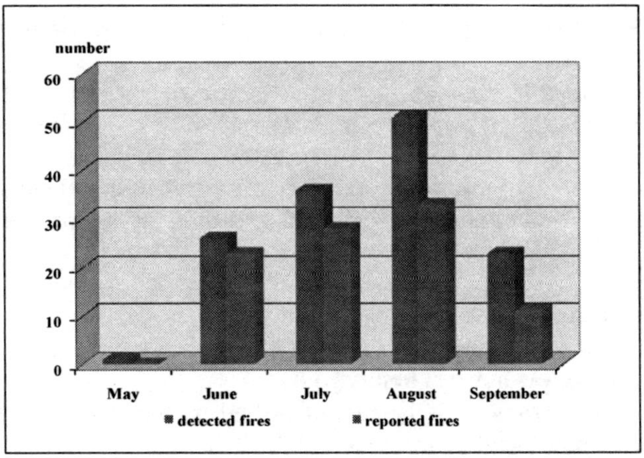

Figure 8: Fire occurrence in months of 2009.

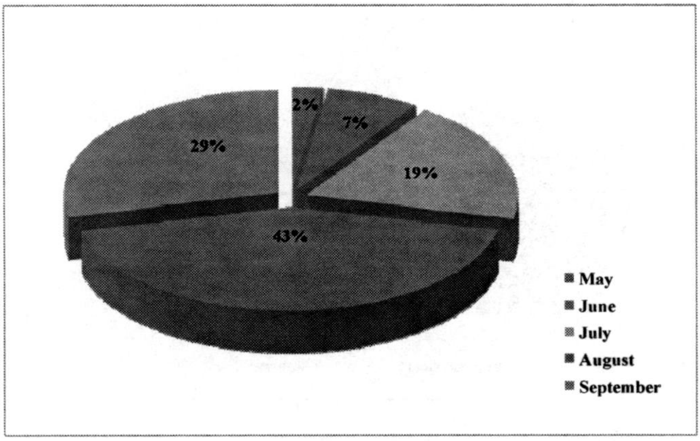

Figure 9: Uncontrolled fires in 2009 given in %.

Controlled (reported) fires were the fires, which were reported to operators of the control centre in advance, and hence, were only monitored or searched for. These kinds of fires were often reported by employees of forest districts, and in most cases these fires referred to prescribed burning of felling waste in forest stands. Uncontrolled fires were the ones that were identified by the system in the control centre, and subsequently the operators performed all relevant measures to suppress them, and to monitor them further.

Overall, during the trial operation the following number of fires was detected:

The highest frequency of uncontrolled fires was, as expected, in the summer months. It is mainly the result of more frequent movement of people in nature during holidays, as well as of favourable natural conditions (drought).

Table 1: Overview of detected fires in the trial years.

Year	Controlled (reported)	Uncontrolled	Total (detected)
2008	139	39	178
2009	95	42	137
Total	234	81	315

Table 2: Overview of detected fires in the trial months.

Month	Controlled (reported)			Uncontrolled			Total (detected)		
	2008	2009	Sum	2008	2009	Sum	2008	2009	Sum
May					1	1		1	1
June		23	23		3	3		26	26
July	47	28	75	11	8	19	58	36	94
August	54	33	87	14	18	32	68	51	119
September	38	11	49	14	12	26	52	23	75
Total	**139**	**95**	**234**	**39**	**42**	**81**	**178**	**137**	**315**

The afternoon hours of workdays between 4 pm and 8 pm, and Saturdays were found to be the most critical time periods for fire occurrence. This can be explained by the occurrence of personal agricultural activities, which are usually carried out by the population in the region in addition to their occupation.

Based on our two-year experience with the operation of the system Forestwatch we can state the following:

4.1 System benefits

- Continuous, automated control of the area of interest,
- Continuous evaluation of the situation in the monitored area,
- Alerting an operator to the changed situation in the area, and indication of the changes of examined state characteristics,
- Displaying information about the cause of alert,
- Determining of indicated problematic area with GPS coordinates and displaying it on a digital map,
- Possibility to define permanent sources of smoke (factory, dwelling – isolated house, …),
- Possibility of manual control of the system – cameras,
- An operator can monitor several areas at the same time – the system is automated, and hence, only the alerts need to be resolved,
- Reduction of costs for monitoring of areas at risk,
- Preventive psychological effect – by publishing information about the monitoring of an area, e.g. in media.

Apart from the listed advantages of direct fire prevention, the system also provides a user with the tools and the possibilities for efficient management and monitoring of forest stands concerning:

- Illegal timber felling
- Illegal movement of persons and motorised vehicles
- Illegal hunting or poaching
- Wildlife movement and migration
- General support of nature protection

4.2 Problems and shortcomings

Strong storms in the region were found to be one of the main, but hardly influence able factors affecting successful operation, as they often caused power outages.

The system itself is sensitive to accurate alignment of directional antennae of Wi-Fi devices, as the whole system is directly dependent on this communication. As far as the installation is precise, the system works reliably.

During the operation of the system, common operational shortcomings were being eliminated, and the whole system was being tuned. Frequently, system breakdown resulted from storms causing power outages on the towers, or from service works and replacement of components of electric power network by the company E-ON (Stredoslovenská energetika SSE – Central Slovak Energetic) in the areas around the towers. Such effects are difficult to be anticipated, with minimum possibility to reduce them by the system operator.

5 Other activities

As the Slovak republic is the first country in Central Europe, and after Greece the second one in Europe that uses the system Forestwatch, its implementation has also attracted interest of partners from abroad. In 2009, reference demonstrations were performed for the interested partners from Latvia, Poland, and Mongolia. The countries were represented by forestry professionals, who were mainly informed about the system implementation, its functionality, and practical experience from the operational trial concerning its functionality in the topographic and climatic conditions of Slovakia. The partners indicated their intentions to implement such a system in cooperation with our experts.

Due to the fact that the system can be applied not only for fire prevention, but also for other purposes (e.g. illegal timber felling, wildlife migration, etc. ...), these aspects were also monitored during the system operation. However, no significant activities or events of such kinds were detected during the trial period.

6 System installation costs and annual operating costs

Since the project was performed as an operational trial, the system was realised by cumulating technological, personnel, and financial sources of the three partners – State Forest Enterprise, National Forest Centre, and the private company ICZ Slovakia. The overall costs for the trial realisation were estimated to be 450,000 € including non-financial inputs.

Average annual costs in the current operational regime are approximately 40 000 €, while both maintenance costs and direct operational costs make each about one half of the total costs.

7 Conclusion

The project of fire monitoring was realised at the Forest Enterprise Kriváň covering approximately 60,000 ha of forests, managed predominantly by the state enterprise Forests of the Slovak republic. During the trial period, the activities dealing with the preparation, realisation, trial operation, and evaluation were performed. The system is functional and suitable for a wider use in the conditions of Slovakia. The fire detection system has been found out to be an efficient complement to ground and aerial monitoring. It is of interest to extend the system to other parts of Slovakia, as this would provide an alternative to the present fire-fighting measures, and in addition, it would also reduce costs, primarily through the establishment of centralised control centres.

Further information

Websites: http://www.eep.eu.com/en/world.html (Europe)

http://www.evsolutions.biz/deployment.php (South Africa)

Semi-expendable Unmanned Aerial Vehicle for forest fire suppression

D. Benavente
Founder, Embention Sistemas Inteligentes, S.L., Spain

Abstract

Rapid detection systems and other surveillance and real-time monitoring systems have become more and more a reality in recent years. The information generated is very useful for avoiding risks and to optimally manage the available firefighting resources. However, there are lots of scenarios for which a rapid and effective response is not possible because of strong winds, the absence of daylight or difficulty in gaining access to a specific location with conventional aerial assets. This paper presents a new forest fire suppression tool capable of handling all these factors: Flamingo (Forest fLAMes INtelliGent Outputter), a low-cost semi-expendable Unmanned Aerial Vehicle (UAV) with a single-use deposit loaded with fire retardants.

Keywords: Flamingo, forest fire suppression technology, civil UAV, guided bomb, severe winds, night operations, Veronte, autopilot.

1 Introduction

In recent years technology has been entering the forest fire world with rapid detection systems, real-time monitoring systems, etc. These systems can indeed contribute to the avoidance of risks and elaborate optimal suppression strategies thanks to the information they provide. What we are presenting here is technology applied to the suppression area.

Our approach tries to fill the gap left by conventional aerial means whose operation is too dangerous in the presence of strong winds and/or during nighttime. This gap has been also targeted by other systems like PCADS [4]. Night operation in particular is one of the key points identified for the near future by many forest fire aerial contractors.

WIT Transactions on Ecology and the Environment, Vol 137, © 2010 WIT Press
www.witpress.com, ISSN 1743-3541 (on-line)
doi:10.2495/FIVA100131

Concerning our approach, the basics of the technology have been used for many years in the past for military purposes. In fact, the way Flamingo works is quite similar to that of any guided bomb being used nowadays. However, we prefer to say that Flamingo is a semi-expendable UAV that mainly consists of 2 parts: a single-use deposit and a re-usable control unit including the fins, servos and autopilot responsible for getting the load of fire retardants right to the coordinates established a couple of seconds before flight.

Although the feasibility of precise self guidance towards a target has been proven with guided bombs unfortunately many times, the cost of this technology might result too high for wildfire suppression. Keeping the overall unit cost well under 1KEUR while keeping flight performances posed a big design challenge on many areas because no fancy materials or precision sensors could be used onboard.

2 System use

In its current version, each Flamingo unit contains 206 litres of fire retardant. Although Flamingo guides itself by deflecting its fins in the air, it is not provided with propulsion means onboard. This means that in order for Flamingo to fly, it needs to be dropped from an aircraft from an altitude of approximately 10000ft. The aircraft we are considering for operation (e.g. CN-235, C-295, C-130 Hercules, A400M, etc) could carry from 20 up to 100 Flamingo units resulting in more than 20000 litres of fire retardant delivered per flight.

The baseline use is as follows: once a forest fire is detected and the system is requested for operation, a cargo aircraft is loaded with a set of Flamingo units. All these units exchange data within a low-power wireless local network. Apart from the set of Flamingo units, the aircraft and its cargo-like launch system onboard, there is still a key part of the system to be mentioned: the Flamingo Air Control Station (FACS) and the Flamingo Ground Control Station (FGCS).

The Flamingo Air Control Station is responsible for transmitting independent target coordinates to each of the Flamingo units onboard. Therefore, FACS also constitutes a node in the wireless local network onboard. In addition, FACS holds a long-range communications link to the FGCS. All the fire information available is displayed to the system operators at both the FACS and FGCS using a GIS-based software tool. In addition, the software allows the user to define a direct attack or an indirect suppression strategy which will result in a set of coordinates to be given individually to each of the Flamingo units inside the aircraft just before launch.

As the aircraft flies towards the fire (at 10000ft altitude), a launch window gets open. At that moment, the set of target coordinates are updated with the latest information available and all the Flamingo units are dropped like cargo. After a couple of seconds after launch the transient gets cancelled and each Flamingo starts to move its fins to get to its target. In order to achieve this, Flamingo is equipped with our autopilot Veronte, which runs the guidance, navigation and control algorithms needed in real time. The navigation algorithm computes at every moment the position, velocity and attitude of Flamingo by making a fusion of GPS, magnetic, pressure and inertial measurements. The guidance algorithm basically ensures that Flamingo can always reach the target at any given moment. Lastly the control algorithm regulates the difference between the estimated variables and the desired ones.

In less than a minute Flamingo hits ground at more than 300km/h breaking the deposit and helping spread the retardant. A couple of seconds before the impact, the control unit gets separated from the deposit right before it deploys a parachute in order to absorb energy itself and avoid a strong impact.

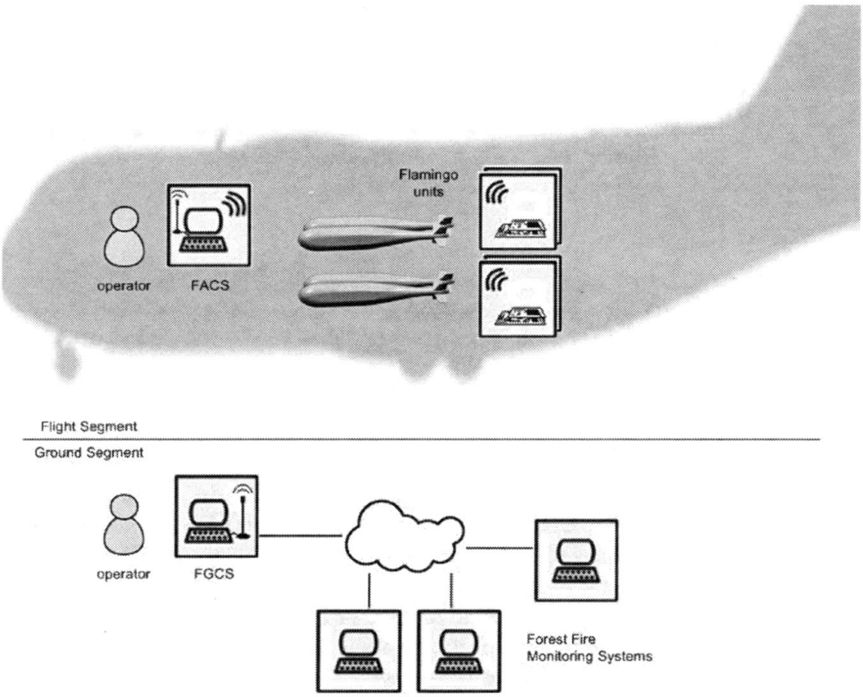

Figure 1: Flamingo UAV architecture.

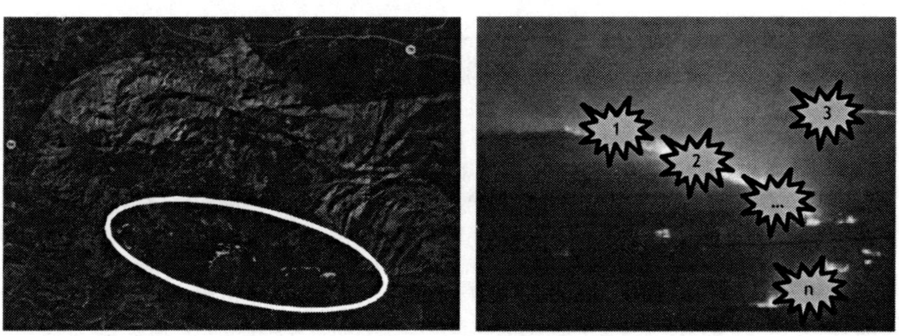

Figure 2: System use: strategy and allocation.

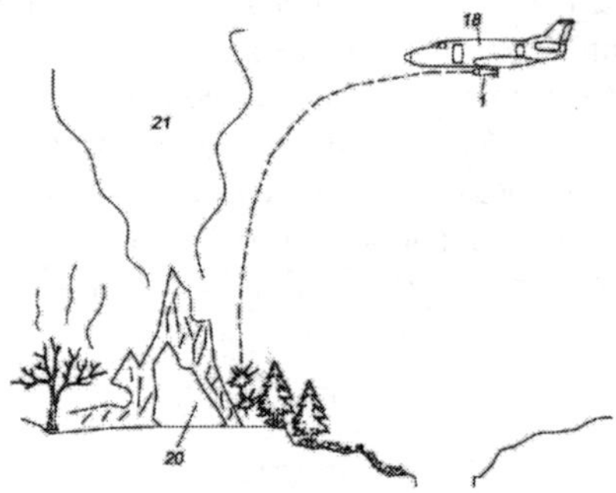

Figure 3: Direct suppression strategy.

3 About Flamingo

The main element of the system presented is Flamingo. As we mentioned earlier we like to say that Flamingo is a semi-expendable UAV composed of 2 main parts: the Suppression Unit and the Control Unit.

3.1 Suppression unit

It is a slender deposit containing the UAV payload, that is, the retardant. Since it is a single-use component it has been designed to minimize its unit cost while its structure is able to withstand a 4.5g manoeuvre while in the aircraft [1]. This component is also the main responsible for inherent aerodynamic stability and controllability. Furthermore, it has to ensure retardant tightness at every time throughout the mission. Minimizing the cost subject to all these constraints was not an easy problem to solve. As of today, the solution we have reached consists of a single piece produced with a rotomoulding technique.

3.2 Control unit

It is the piece adding the intelligence and control capabilities to Flamingo. It contains a set of aerodynamic surfaces deflected by a set of servos via a transmission system. These fins are moved by our Veronte autopilot as an outcome of our embedded algorithms which take into account not only the target coordinates but also the measurements from its onboard mounted sensors: GPS, inertial measurement unit, magnetometer, and air data system. Veronte is also responsible for separating the control unit from the deposit at the right time and deploying a parachute right away to avoid a strong impact. It is important to mention that the control unit by itself is also a structural piece because it has to transmit all aerodynamic loads from the fins to the deposit.

Figure 4: Suppression unit and control unit.

Figure 5: Frames from 5-liter prototype impact on 20m x 20m concrete platform.

Figure 6: Beta series ready to fly.

4 Current status

After a conceptual 5-liter prototype and a beta series of 60 litres, we developed our product scaled on a 206-liter deposit. Both the suppression and control units were carefully designed taking into account the lessons learnt with the prototypes and the beta series.

At this moment the product Flamingo is involved in a qualification process for aeronautic products at INTA (Instituto Nacional de Técnica Aeroespacial, Spain) for its integration and operation with the EADS CASA C-212 aircraft using a cargo door drop system [2].

As part of the qualification process, full system tests will start shortly with the collaboration of INTA, where the extinguishing power of each deposit will constitute one of the main results. One of the main difficulties we have faced so far is the complexity to carry out tests. Not only it is not easy to obtain burn permits in Spain but also it is extremely difficult to get approvals for experimental bombs to be dropped from an airplane. However, since it was one of the main challenges in our project we managed to test our autopilot and control system in flight with our 5-litre prototypes from 1800ft.

References

[1] A. Bedmar, D. Benavente, M. Breen, M. García, J. González, D. Gutiérrez, *Flamingo Technical Specification*, david.benavente@embention.com
[2] L. Dávila, R. Dorado, B. Marqués, AER/PRO/7130/003/INTA
[3] RD 2218/2004, Reglamento de Aeronavegabilidad de la Defensa (RAD)
[4] PCADS, http://www.caylym.com/
[5] MIL-HDBK-1791 (USAF) *Designing for Internal Aerial Delivery in Fixed-Wing Aircraft*

Meteorological condition and numerical simulation of the atmospheric transport of pollution emitted by vegetation fires

A. M. Ramos[1,2], F. C. Conde[1,2], S. Freitas[3], K. Longo[3], A. M. Silva[2], D. S. Moreira[3], P. S. Lucio[1,4] & A. L. Fazenda[3]
[1]*Geophysics Centre of Évora (CGE), University of Évora, Portugal*
[2]*Instituto Nacional de Meteorologia (INMET), Brasília-DF, Brazil*
[3]*Center for Weather Prediction and Climate Studies – CPTEC/INPE, Brazil*
[4]*Departamento de Estatística (DES), Universidade Federal do Rio Grande do Norte, Brazil*

Abstract

The objective of this study is to investigate the atmospheric transport of gases and particles emitted by forest fires occurring on the Iberian Peninsula, affecting Continental Portugal during the period from 7 to 12 August 2003. The simulations were implemented using the on-line 3-D transport model CATT-BRAMS (**C**oupled **A**erosol and **T**racer **T**ransport to the **B**razilian developments on the **R**egional **A**tmospheric **M**odeling **S**ystem) coupled to an emission model. The results generated by CATT-BRAMS allow one to describe the local and synoptical condition at the target area. The wind direction from the northeast varying to east over the Iberian Peninsula favored the dislodgment of the smoke plume toward the Atlantic Ocean, distant from the regions with forest fire emissions.

Keywords: atmospheric modeling, biomass burning, summer 2003, long-distance transport.

1 Introduction

Biomass burning is a major source of regional and global scale air pollution, and the smoke plumes interact with both solar and terrestrial radiation, sometimes

WIT Transactions on Ecology and the Environment, Vol 137, © 2010 WIT Press
www.witpress.com, ISSN 1743-3541 (on-line)
doi:10.2495/FIVA100141

exerting significant regional-scale forcing of climate. Wildfire initiation and spread are known to be heavily influence by wind (Clements *et al* [1]). Direction spread, propagation speed and smoke plume are result of strong interactions between wildfire and atmosphere occurring at different scales (turbulent mixing in the front, large eddies forming near the front, fire induced winds). Fire emissions are mainly concentrated in the tropics, and are an important regulator of the oxidizing capacity of the global atmosphere. To move forward, fire science needs to go beyond observation and description to provide increased understanding of the interactions between fire and forests so that effective policy, management, modeling and monitoring efforts can be created.

Through this route, biomass burning affects the global budgets of many chemistry species, including the second most important emitted greenhouse gas, methane. There are also important post-fire effects on soil and vegetation emissions. A variety of carbon and nitrogen species are released into the atmosphere from vegetation fires. Carbon monoxide, methane and nitrogen oxides lead to the photochemical production of ozone (O_3) in the troposphere (Crutzen [2]). Particulate material, such as smoke or soot particles, are also produced during the burning process and released into the atmosphere. These solid particles absorb and scatter incoming sunlight and hence impact the local, regional and global climate. In addition, particulates with diameter of about 2.5 micrometers or smaller can lead to various human respiratory and general health problems when inhaled.

The objective of this work is to study the atmospheric transport of gases and particles emitted by biomass burning, due to forest fires affecting large areas over the European continent (Iberian Peninsula and France), during the occurrence of heat wave that affected Europe and in particular continental Portugal on the period from 7 to 12 August 2003. The synoptical situation responsible of the heat wave event was also studied. These studies were accomplished through the use of numerical simulations with the on-line 3-D transport model CATT-BRAMS (**C**oupled **A**erosol and **T**racer **T**ransport to the **B**razilian developments on the **R**egional **A**tmospheric **M**odeling **S**ystem) coupled to an emission model (Freitas *et al* [3]).

2 Methods

2.1 Fires description

According to the European Forest Fires Information System [4], during the 2003 summer heat wave in Europe over 25,000 fires in Portugal, Spain, Italy, France, Austria, Finland, Denmark and Ireland were observed. The total area of forest burnt was 647,069 hectares – four times the size of Greater London. More than half (390,146 hectares) were in Portugal, making it the worst forest fire season the country had faced in the last 23 years with a total area burned of almost five times the average area and an impressive amount of natural resources affected. Spain registered during the same period extreme temperatures of 46°C in the south and 51°C in the city of Sevilla. Forest fires burning 70 kilometers east of

Madrid and in Salamanca destroyed about 2.000 hectares of forest. Southern France saw its share of forest fires in the region of the Var, near the Mediterranean Coast and in the Tarn.

Portugal's forest area is about 3.3 million hectare, of which 87% corresponds to private property. In terms of the monthly distribution the year of 2003 was characterized by a remarkable concentration of burned area in August, where 66% of total burned area took place, which is about 2.7 times greater than the annual average value of the last 10 years. It was also in this month that the highest number of fires was registered, according to the authorities; forest fires have affected 15 of the country's 18 districts, with most of fires occurred in the Northern Region (69%).

This region is characterized by a large density of population on the forestland, associated with small dimension land properties. According to the European Forest Fire Damage Assessment System (EFFDAS) evaluation the distribution of the forest fires covering areas larger or equal than 50 ha recorded in 2003 in Portugal represented only 2% of the total number of fires (327 fires) but responsible for 95% of the total burned area. The higher number of fires occurred mostly over Guarda district representing 32% of the total incidents (see table 1). However the higher values of burned area were in the districts of Castelo

Table 1: Total burned area as distributed through the Portuguese administrative districts in August 2003. The districts most strongly affected by the fires were Castelo Branco (21% burned area), Portalegre (16%), Santarém (15%) and Faro (14%), August 2003.

Start date	Districts	Area burned (ha)
07/08/2003	Bragança	1677
	Guarda	10842
	Castelo Branco	892
	Leiria	1785
	Faro	25900
08/08/2003	Bragança	1376
	Portalegre	2530
09/08/2003	Faro	950
	Guarda	700
10/08/2003	Guarda	700
12/08/2003	Guarda	3618
	Castelo Branco	1560
	Faro	14850

Branco and Portalegre, where 10% belongs to agriculture areas mainly located in Alentejo, of which Portalegre represents 37% of the total.

In Portugal, there were 21 deaths during 2003 fire season, 18 of them only in three weeks (between July 29 and August 14). Most of the reported deaths were located in the centre region of Portugal (districts of Castelo Branco, Portalegre and Santarém) and occurred during the critical period where several forest fires created a continuous burned area. Most of the victims (17) were civilians trying to save goods or escape from fire. There were also reports of more than one thousand people (mainly civilians) needing medical assistance due to smoke intoxications, burns and wounds. Total amount of damages has achieved a value of more than 1000 M€. Over 2.000 buildings were affected, causing almost 200 homeless. Additionally, more than two thousand km of electrical cables were destroyed, leaving over half- million people without electricity. Telephone network was also destroyed in some areas, leading to absence of communication with more than 10 thousand homes (National Service for Fire and Civil Protection and Technological Hazards Division).

2.2 Description CATT-BRAMS model

The on-line 3-D transport model follows the Eulerian approach and was coupled to the Brazilian developments on the Regional Atmospheric Modeling System (B-RAMS). The RAMS is a multipurpose, numerical prediction model designed to simulate atmospheric circulations spanning in scale from hemispheric scales down to large eddy simulations (LES) of the planetary boundary layer. The equation set used is the quasi-Boussinesq nonhydrostatic equations described by Tripoli *et al* [5]. The model is equipped with a multiple grid nesting scheme which allows the model equations to be solved simultaneously on any number of interacting computational meshes of differing spatial resolution. It has a complex set of packages to simulate processes such as: radiative transfer, surface-air water, heat and momentum exchanges, turbulent planetary boundary layer transport and cloud microphysics. The initial conditions can be defined from various observational data sets that can be combined and processed with a mesoscale isentropic data analysis package (Tremback [6]). For the boundary conditions, the 4DDA schemes allow the atmospheric fields to be nudged towards the large-scale data. New deep and shallow convective schemes based on the mass flux approach and with several types of closure (Grell and Devenyi [7]) were also implemented.

The CATT-BRAMS explores the B-RAMS tracer transport capability using slots for scalars. The on-line transport model solves the mass conservation equation for carbon monoxide (CO) and particulate material $PM_{2.5}$, considering a tracer mixing ratio, $\boldsymbol{s}$ ($=\rho/\rho_{air}$) where ρ stands for the concentration of the tracer. This equation is solved taking in account the following processes:

$$\frac{\partial s}{\partial t} = \left(\frac{\partial s}{\partial t}\right)_{adv} + \left(\frac{\partial s}{\partial t}\right)_{\substack{PBL\\turb}} + \left(\frac{\partial s}{\partial t}\right)_{\substack{deep\\conv}} + \left(\frac{\partial s}{\partial t}\right)_{\substack{shallow\\conv}} + W_{PM\,2.5} + R + Q \quad (1)$$

where ***adv, PBL turb*** and ***deep (shallow) conv*** stand for grid-scale advection, sub-grid transport in the planetary boundary layer (PBL) and sub-grid transport associated to moist and deep (shallow) convection, respectively. ***W*** accounts for the convective wet removal for $PM_{2.5}$, ***R*** is a sink term associated with generic process of removal/transformation of tracers (dry deposition for $PM_{2.5}$ and chemical transformation for CO), and ***Q*** is the source emission associated to the biomass burning process.

The tracer mixing ratio ***s*** is updated in time using the total tendency given by Equation (1) and a constant inflow is applied as a tracer boundary condition. A complete description of the terms of the Equation 1 is given in Freitas *et al* [3]. The advection and PBL turbulent transport schemes are obtained from the B-RAMS. The sub-grid transport associated with deep and shallow convective transport is coupled to the Grell cumulus scheme.

The biomass burning source emission parameterization (for CO, CO2, CH4, NO_x and $PM_{2.5}$) is based on the MODIS (Moderate Resolution Imaging Spectroradiometer) fire as well as field observations. For each fire detected by remote sensing, the mass of emitted tracer is calculated and its emission in the model follows a diurnal cycle of the burning (Freitas *et al* [3]). The type of vegetation that is burning is obtained from the IGBP 1km vegetation map, thus allowing an appropriate selection of the vegetation dependent factors in Equation 1.

The sources are spatially and temporally distributed and daily assimilated according to the biomass burning spots defined by the satellite observations. The carbon monoxide emission associated to the anthropogenic processes (industrial, power generation, transportation, etc) is provided by EDGAR database (http://arch.rivm.nl/env/int/coredata/edgar/). All biomass burning emissions are added with the EDGAR "agricultural waste burn" and "fuelwood burning" emissions with 1x1 degree horizontal resolution and 1 year time resolution. For $PM_{2.5}$, the tracer convective transport scheme accounts also for the wet (in and below cloud) deposition based on the work of Berge [8].

2.2.1 Model configuration, initial and boundary conditions

The main options and parameterizations used in these simulations, for our goal, are described as follows:

- For this study the initial and boundary conditions necessary to drive CATT-BRAMS were provided by the twice daily Aviation run of the National Centers for Environmental Prediction Global Spectral Model (AVN) with a resolution 1.25 x 1.25 degrees. The analyzed fields include geopotential, temperature, wind (u,v), and relative humidity at 11 isobaric levels, and the surface pressure every 6 hours intervals (00, 06, 12 e 18 UTC);
- The model is set up with three tri-dimensional grids. The coarse grid specification was defined with 80 km grid spacing, the other two with 20 km and 5 km horizontal resolutions respectively, both centered at 38.8°N; 9.28°W (Lisbon, with altitude reaching 104 m a.s.l). The vertical resolution starts at 150 m near the surface, stretching at a rate of 1.10 to a final resolution of 850 m, with the model top at about 21 km. The coarse grid

covers Europe, Atlantic Ocean and North Africa. The simulation covered 91 days, beginning July 01 at 00:00 UTC;

- The B-RAMS full microphysics package was activated for all the grids. This scheme includes the use of generalized gamma distributions as the basic function for all hydrometeor species; the use of a heat budget equation for hydrometeor classes, allowing heat storage and mixed phase hydrometeors; partitioning hydrometeors into seven classes (including separate graupel and hail categories); the use of stochastic collection rather than continuous accretion approximations and the extension of the ice nucleation scheme to include homogeneous nucleation of ice from haze particles and cloud droplets (Walko *et al* [9]);
- Grell cumulus parameterization scheme improved by Grell and Devenyi [7], radiation parameterization from Chen and Cotton [10], turbulence and diffusion parameterizations were handled using the Mellor and Yamada [11] was activated in three grids;
- Topography, vegetation type, land percentage and sea surface temperature were read onto the grid from USGS (U.S. Geological Survey at 1km horizontal resolution) datasets. The simulation utilized silhouette-averaged topography (Bossert [12]) in order to incorporate the desired terrain effects.
- A soil model was assumed, using seven levels with 50% saturation moisture for all depths (Tremback and Kessler [13]) and constant inflow conditions are used to the tracer boundary condition.

3 Results

Meteorological conditions: The southeast region of Iberian Peninsula Southeast registered extreme conditions on August 2003, with very high temperatures and unstable atmosphere which favoured fire raged out of control through several parts of Spain and Portugal. The synoptic configuration during the period verified a wide anticyclone system over North Atlantic Ocean as the averaged streamlines at 850 hPa (fig.1(a)) and at 200 hPa (fig. 1(b)) shows, characterizing atmospheric circulation associated with blocking anticyclone. The wind is practically zonal in the south flank of the anticyclone at 200 hPa extending through a quite considerable longitudinal area, with the persistence of the anticyclone over North Atlantic Ocean, which confirms the stationary location of blocking. This anomalous anticyclone is also seen in the anomalies of geopotential height at 500 hPa under the form of an extensive area of positive anomalies (fig. 1(c)). Situation of negative anomaly of relative humidity (fig. 1(d)) is observed in the whole region with positive temperature anomalies extending from Atlantic Ocean to Iberian Peninsula (fig. 1(e)) with negative core pressure anomalies in this area (fig. 1(f)). The persistent anticyclonic conditions characterized by exceptionally high temperatures low values of relative humidity, mainly for the first two weeks of August 2003, with an average flow primary from northwest over Iberian Peninsula, had a significant impact in what respects to the number of forest fires and extension of the burned areas and the development of a large-scale pollution episode.

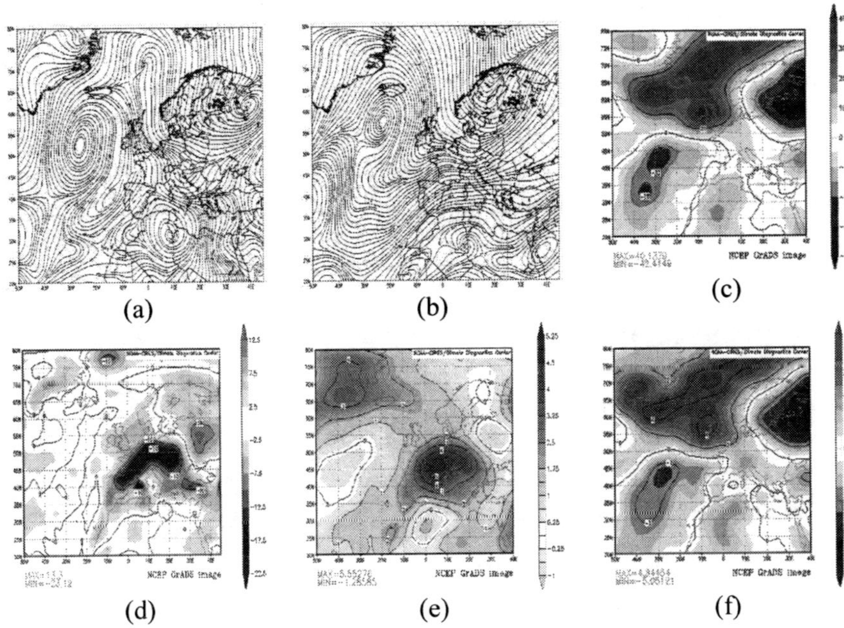

Figure 1: Averaged streamlines at 850 hPa (a) and at 200 hPa (b), geopotential height anomalies (c) at 500 hPa; relative humidity (d) and temperature (e) at 1000 hPa and atmospheric pressure (f) at 850 hPa, August 2003.

Simulation of smoke plume: The analysis correspond to the average carbon monoxide (CO) concentration values (ppb) at two different depths, 53 and 1100 meters, above the surface and of the total column content $PM_{2.5}$ ($\mu g.m^{-2}$) aerosols during the period 7–12 August 2003. The wind fields for the three tri-dimensional grids: 80 km (grid 1), 20 km (grid 2) and 5km (grid 3) are used in the previous simulations. The transport is conditioned by two anticyclones one located over the British Islands and the other one over Azores involving practically all grid domains. Both originate a wind confluence zone along the European continent and North Africa. The wind over Iberian Peninsula, varying from northeast to the east, influences the transport smoke plume. Hence the transport is firstly across the Portuguese continent from Spain, where forest fires occur at different places and then there is a predominant drainage of a part plume towards the south Atlantic Ocean leaving the continent around latitude 37°30'N. A part smoke plume from Galiza in Spain and from north Portugal is transported directly towards the Atlantic Ocean due to the influence of meridional wind flow. An anticyclone over Mediterranean Sea conditions the export of the smoke plume observed over Africa and east Europe. A part plume is displaced towards the southeast of the European continent, including the Mediterranean area, whereas the other part is advected over North Africa mixing with the plume and transported to the Atlantic Ocean. A part smoke plume from France moved to Bay of Biscay and then is redirected to the Atlantic Ocean (fig. 2).

The use grid 2 allows the nesting-grid to higher-resolution simulations of the CO and $PM_{2.5}$ concentrations (grid 3) over Iberian Peninsula. The CO concentrations detailed over the continent show values about 3000 ppb and a maximum core of 5000 ppb mainly in north and centre-south regions of Portugal at both altitudes. In Spain, the maximum values are located along the northern regions as well as at Extremadura, Andalusia and Madrid regions with concentrations of about 100 ppb. The transport is conditioned by two

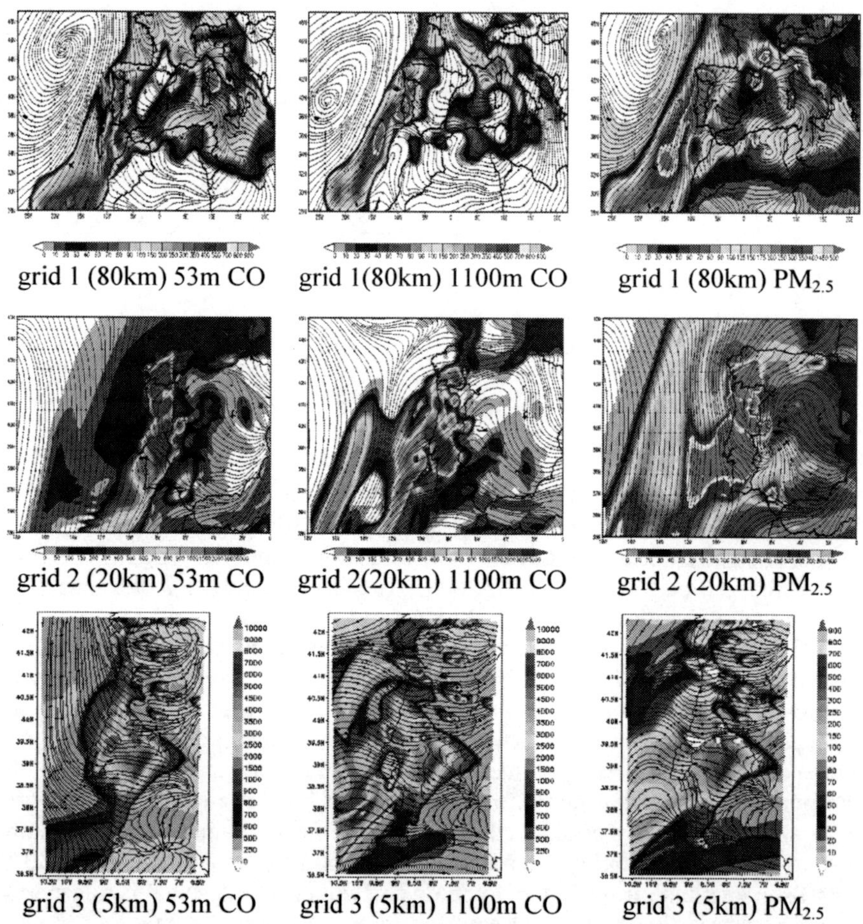

Figure 2: The average smoke plume values of the $PM_{2.5}$ total column content in ($\mu g.m^{-2}$) and CO concentrations in parts per billion (ppb) at two different altitudes , 53 and 1100 meters, with wind field simulated by CATT-BRAMS for grids 1 (80km), 2 (20 km) and 3 (5km), during the period 7–12 August 2003.

anticyclones one located over Atlantic Ocean and the other one over Mediterranean Sea with exports predominantly smoke plume over the Atlantic Ocean reaching the proximity African continent. For grid 3 (5 km) the transport from the fires in the Portugal was influenced by oscillation wind confluence zone over the country, causing two main different exports patterns of plume, one for the interior continent reaching bordering Spain and other to the Atlantic Ocean.

Fig. 3 shows the times series of simulated CO concentrations during the period 7–12 August 2003, for different districts in Portugal, due to forest fires classified according to EFFDAS system, using the higher-resolution wind simulations grid 3 (5km). The CO evolution within planetary boundary layer

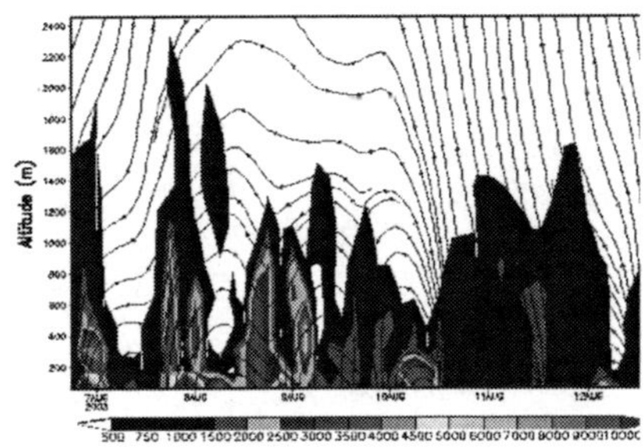

(a) Bragança

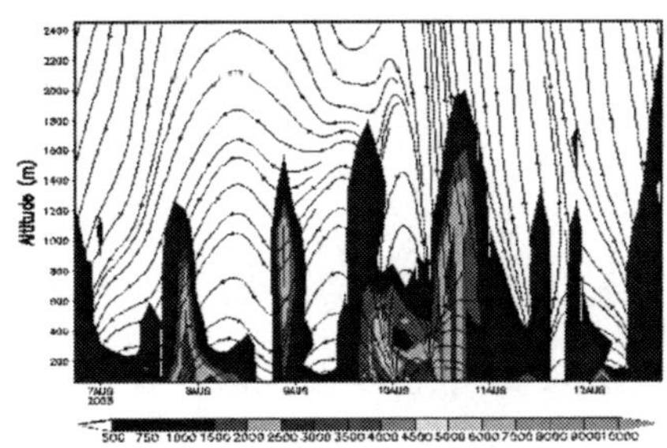

(b) Guarda

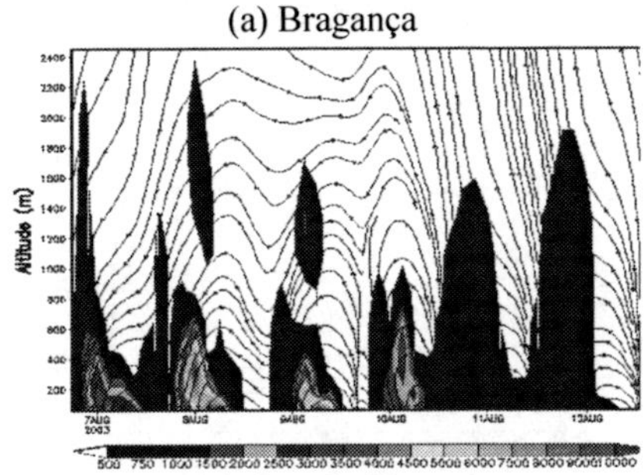

(c) Castelo Branco

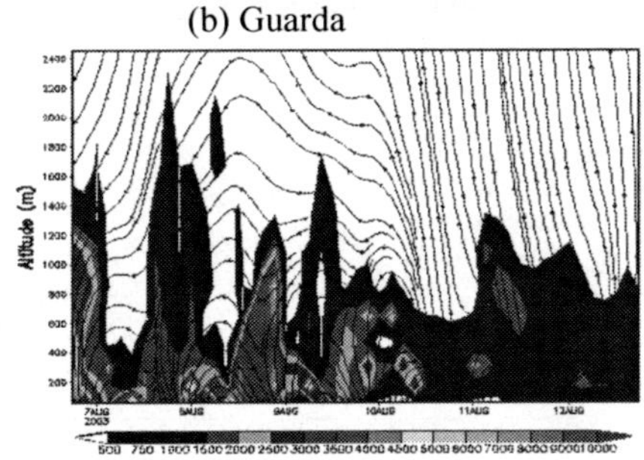

(d) Leiria

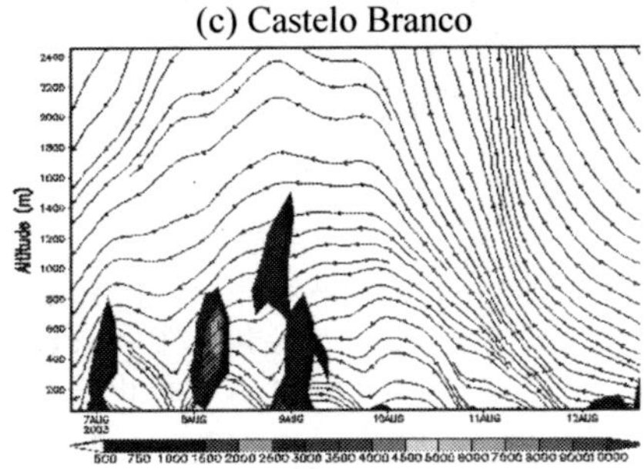

(e) Portoalegre

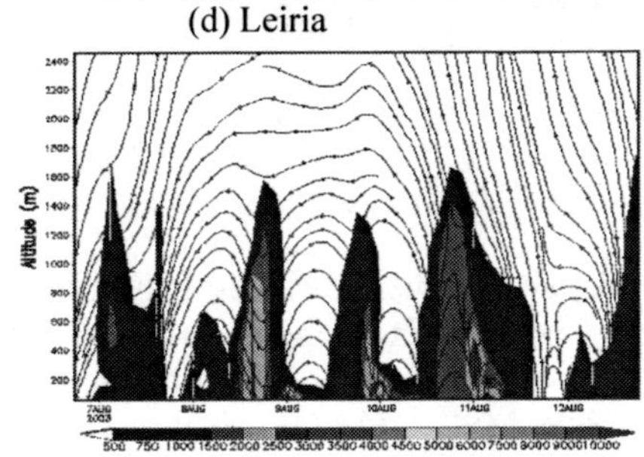

(f) Faro

Figure 3: The time series of concentration monoxide carbon (CO), in ppb, with wind field simulated by CATT-BRAMS for grid 3 (5 km), during the period 7–12 August 2003 to districts in Portugal.

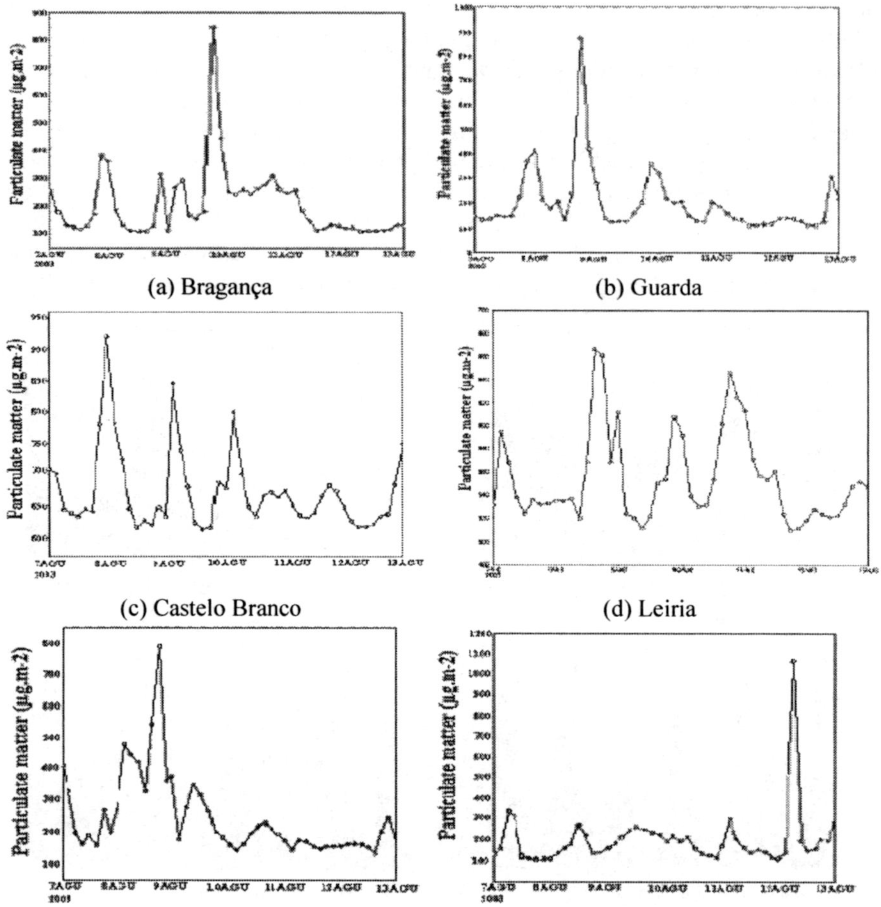

(a) Bragança (b) Guarda

(c) Castelo Branco (d) Leiria

(e) Portalegre (f) Faro

Figure 4: The time series of particulate matter ($PM_{2.5}$), in $\mu g.m^{-2}$, simulated by CATT-BRAMS for grid 3 (5 km), during the period 7–12 August 2003.

(PBL) shows, in all districts, values of about 4.500 ppb confined to an altitude of approximately 2000 meters, with the exception of Portalegre where the values are confined up to 1500 meters. The gas is transported vertically and homogenized inside the mixture layer decreasing in intensity. The export and expansion of the pollutant in horizontal direction and relight the fires (as described in the EFFDAS) can provide high concentrations of pollutant in the atmosphere. The times series simulated particulate matter $P.M_{2.5}$ concentration profile in $\mu g.m^{-2}$ is shown in Fig 4. The modeled results showed maximum concentration values of 950 $\mu g.m^{-2}$ and are obtained mainly for the forest fire event of 09/08/2003. The Faro district shows a maximum on the 12/08/2003 corresponding to a series of forest fire events of great intensity according to the

Forest Portuguese Authorities well as Castelo Branco with higher value record during period.

4 Conclusions

The numeric simulations of the atmospheric transport of gases and particles emitted by forest fires during the occurrence of a heat wave that affected Continental Portugal on the period from 7 to 12 August 2003 were performed using the on-line 3-D transport model CATT-BRAMS (**C**oupled **A**erosol and **T**racer **T**ransport to the **B**razilian developments on the **R**egional **A**tmospheric **M**odeling **S**ystem) coupled to an emission model, which explores the B-RAMS tracer transport capability of using slots for scalars. It is an on- line transport model fully consistent with the simulated atmospheric dynamics. The results show the large-scale circulation may be responsible for transporting of smoke plume to distant regions away from the burned sites and covering thousands of square kilometers. The general performance of the model simulations lead to believe that the mesoscale models are a useful tool to describe the atmospheric transport of pollutants over the region. Moreover, the long time permanence of adverse meteorological conditions conjugated in some areas with a high vegetable cover density and a pronounced irregular topography as occurring in the north and center of continental Portugal, have contributed to live sceneries (high temperatures and low relative humidity of air characterizing he environment hot and dry). The biomass combustion emits gases and aerosol particles that interact efficiently with the solar radiation affecting the microphysical processes, dynamics of cloud formation and air quality. The contamination caused by fires can reach distant areas from the burned region and increase the atmospheric pollution of urban and industrial areas. Therefore, the understanding of the evaluation and progress of fires, through numeric modeling simulations contributes to identify the interrelations between biosphere and atmosphere and help authorities to prevent human lost.

References

[1] Clements, C. B., Potter, B.E., Zhong, S. In situ Measurements of Water Vapor, Heat and CO2 Fluxes within a prescribed Grass Fire. *International Journal of wildland Fires*, 15(3), 299-306, 2006.

[2] Crutzen, P. J. Tropospheric Ozone: An Overview, Tropospheric Ozone. Edited by Isahsen, IS., D. Reidel, Norwell, Mass., 3-32, 1988.

[3] Freitas, S., K. Longo, M. Silva Dias, P. Silva Dias, R. Chatfield, E. Prins, P. Artaxo, G. Grell and F. Recuero. Monitoring the transport of biomass burning emissions in South America, *Environmental Fluid Mechanics*, *5 (1-2)*, 135 – 167, doi: 10.1007/s10652-005-0243-7, 2005.

[4] Schmuck, G., San-Miguel-Ayanz, J., Barbosa, P., Camia, A., Kucera, J., Libertà, G., Bucella, P., Schulte E., Flies, R., Colletti, L.. Forest Fires in Europe – 2003 Fire Campaign, Official Publication of the European Communities, SPI.04.124.EN, 2004.

[5] Tripoli, G. J., Cotton, W. R.The Colorado State University three-dimensional cloud mesoscale model, 1982: PartI: General theoretical framework and sensitivity experiments, *J. de Rech. Atmos.*, *16*, 185-220, 1982.
[6] Tremback, C. J. Numerical simulation of a mesoscale convective complex model development and numerical results, *Atmos. Sci. 465*, Department of Atmospheric Science, Colorado State University, Fort Collins, CO 80523, 1990.
[7] Grell, G. A., and Devenyi, D. A generalized approach to parameterizing convection combining ensemble and data assimilation techniques, *Geophy. Res. Let.*, *29*, no. 14, 2002.
[8] Berge, E. Coupling of wet scavenging of sulphur to clouds in a numerical weather prediction model, *Tellus*, *45B*, 1-22, 1993.
[9] Walko, R. L., Cotton, W. R., Meyers, M. P., Harrington, J. Y. New RAMS cloud microphysics parameterization. Part I: the single-moment scheme, *Atmos. Res., 38*, 29-62, 1995.
[10] Chen, C. and Cotton, W. R. A one dimensional simulation of the stratocumulus capped mixed layer, *Bound-Layer Meteor.*, *25*, 289-321, 1983.
[11] Mellor, G.L., T. Yamada. A hierarchy of turbulence closure models for planetary boundary layers, *J. Atmos. Sc*i., *31*, 1791-1806, 1974.
[12] Bossert, J. E. Regional scale flows in complex terrain: An observation and numerical investigation, *Paper N° 472*, 254 pp., Department of Atmospheric Science, Colorado State University, Colorado, U.S.A, 1990.
[13] Tremback, C. J., Kessler, R. A surface temperature and moisture parameterization for use in mesoscale numerical models, 7th AMS Conference on Numerical Weather Prediction, Montreal, Canada, *Amer. Meteor. Soc.*, Boston, 355-358, 1985.

Section 4
Decision support systems

SIRIO high performance decision support system for wildfire fighting in alpine regions: an integrated system for risk forecasting and monitoring

L. Corgnati, A. Losso & G. Perona
Department of Electronics, Politecnicodi Torino, Italy

Abstract

In order to optimize performances and resources operating in wildfire fighting in complex orography regions, **SIRIO**, an integrated monitoring system for high performance decision support has been developed. SIRIO is built up of four modules: risk forecasting, monitoring activities, image interpretation and geo-referencing and decision support products generation. The previsional evaluation of the fire risk has an addressing function on surveillance and monitoring activities. The monitoring system operates with low cost optical sensors scanning Visible, Near Infrared and Thermal Infrared bands and a high precision low cost moving system. The system is equipped with a Micro Weather Radar operating as a super gauge and is able to nowcast weather conditions. Processed data are transferred to the central server via a very flexible communication system that can operate with GPRS standards, RF links or satellite connectivity. The image interpretation modules operate on the central server performing smoke detection and hot spot identification on the basis of a tailored radiometric model. The last block of the system is responsible for processed images geo-referencing and decision support products generation. The final products are a collection of scenario images sensed in the Visible band, geo-referenced images with highlighted alarm pixels with overlaying Digital Elevation Model (DEM) levels and topographic layers containing information to be used in the case of intervention, namely a field of view on the analysed area, hot-spot positions, helicopter landing spot positions, water supply positions and intervention squad localization. Monitoring sessions can be browsed on the official SIRIO website, which allows selected access for competent operators.

WIT Transactions on Ecology and the Environment, Vol 137, © 2010 WIT Press
www.witpress.com, ISSN 1743-3541 (on-line)
doi:10.2495/FIVA100151

SIRIO is a valuable aid in fire fighting management, allowing the involved agencies an efficient resources handling (both logistic and human), finalized territory monitoring and intervention planning oriented to operators' safety.
Keywords: forest fires, decision support, monitoring, alpine regions, fire risk.

1 Introduction

The probability of fires rising in the forested environment is steadily increasing owing to climate changes and human activities. Wildland fires are a very prevalent disturbance in the global landscape, causing many serious negative impacts on human safety, health, regional economies and global climate change, with several hundred million hectares of vegetation burning every year. In particular, forest fires in alpine regions are even more dramatic. Complex orography environments are characterized by high spatial variability of physical parameters, hard environmental and weather conditions for monitoring hardware and efficiency and by accessibility problems strongly limiting intervention activities and damage assessment. In order to optimize performances and resources operating in this framework, the Remote Sensing Group (RSG) of Politecnicodi Torino developed SIRIO, an integrated monitoring system for high performance decision support. SIRIO is built up of four modules: operating the risk forecasting (performed by **FIRECAST®**), monitoring activities, image interpretation and geo-referencing and decision support products generation. Combined fire risk simulation and fire statistic computation over the investigated region allow the system to automatically select critical areas to be monitored. Furthermore, the previsional evaluation of the fire risk has an addressing function on surveillance and monitoring activities. The monitoring system operates with low cost, multispectral optical sensors scanning Visible, Near Infrared and Thermal Infrared bands mounted on a high precision, high endurance, low cost moving system, equipped with IP68 sensors cases. The system is equipped with a weather station, in order to collect ancillary data concerning air temperature, air relative humidity, atmospheric pressure, rainfall amount and wind speed and direction. The weather data section is completed by micro weather radar, the **MicroRadarNet (MRN) SuperGauge®**, which is able to detect the actual rainfall from the cloud set and compute rainfall nowcasting. Monitoring scans and schedules, data acquisition, panoramic image composition and data transfer on a central server are managed by the **VM95® controller**, a high performance, low cost, low consumption; high flexibility control system. The sensed images, after being processed by VM95®, are transferred to the central server via a low cost, high performance, high flexibility communication system that can operate with GPRS standards, RF links or satellite connectivity, according to the location and coverage requirements. The image interpretation modules operate on the central server. Images transmitted by the monitoring stations are processed by the **Smoke Detection System**, a powerful tool performing a chromaticity analysis on images for a first step 'static' smoke plumes detection and a feature moving correlation on critical images in order to achieve a second step 'dynamic' smoke detection, which allow for the reduction of false alarms. The **Hot Spot Identification Tool** analyses images within a radiometric model and implements a tailored progressive thresholds system applied to a combination of different frequency band (Visible, Near Infrared, Thermal Infrared) images related to the same scenarios. The system is capable of the identification and localization of fire hot-spot pixels. The thresholds system is customizable by the user, according to the particular characters of the monitored territory. **The Decision Support System** is responsible for the geo-referencing of the processed images and for the generation of decision support products. The final products are a collection of scenario images sensed in the Visible band, geo-referenced images with highlighted alarm pixels with overlaying DEM levels and topographic layers containing information to be used in the case of intervention. Final images contain geo-referenced information about sensors' field of view on the analysed area, hot-spot positions, helicopter landing spot positions, water supply positions and intervention squad localization. The processed images and final products related to active monitoring sessions can be browsed on the official SIRIO website (www.incendiboschivi.com), which allows selected access for competent operators.

2 The system devices

The monitoring platform implemented in SIRIO is the Conway C995 moving system, which guarantees high performances in precision, reliability, long endurance and consumption. The moving engine operates pan and tilt movements with 0.1° precision. Monitoring sensors work onboard the moving system, protected inside IP68 cases. The cases are equipped with Gallium lenses in order to optimize the performances of the Thermal Infrared (TIR) sensor, avoiding flare occurrences on the lens. SIRIO operates a multispectral scan in Visible (VIS), Near Infrared (NIR) and Thermal Infrared bands. The TIR sensing is performed with a common thermal camera. Taking advantage of the sensitivity of CCDs ([400÷1200] nm) in both VIS and NIR bands, SIRIO mounts commercial sensors for the VIS and NIR monitoring. After the removal of the inner CC1 filters from the common photo cameras and video cameras, external CC1 and IR filters are applied to the sensors for the selection of the sensing frequency band. In order to perform a monitoring activity, taking into account the territory orography, video cameras and photo cameras are equipped with zoom lenses. The VM95® controller manages the power supply for the whole system, the moving system, the communication system, the weather station, the sensors settings, the scans, the image acquisition and composition, the data flows, the data storage, the fire risk forecasting system, the image interpretation tools, the decision support system and the user controls. VM95® operates inside an IP68 case and is equipped with solar panels and an emergency battery pack. The VM95® controller is a high performance, low cost, low consumption, high flexibility control system. VM95® is remotely programmable for automatic sessions and the user can remotely control it for real-time off-schedule scanning and statistical analyses. Small dimensions, low cost, low computational needs and low consumption features allow the VM95® to be installed onboard the monitoring stations. Processed data are transferred to the central server via a low cost, high performance, high flexibility communication system that can operate with GPRS standards, RF links or satellite connectivity, according to the location, coverage and bandwidth requirements. SIRIO collects ancillary weather data concerning air temperature, air relative humidity, atmospheric pressure, rainfall amount and wind speed and direction. These data are used in the evaluation of fire statistics and risk forecasting. Furthermore, weather data are crucial in the management of intervention activities and in the generation of decision support products. The weather section is completed by MRN SuperGauge®, a low cost short-range X band micro weather radar that is able to evaluate rain fields, detect the actual rainfall from the cloud set and compute rainfall nowcasting.

3 The system architecture

SIRIO optimizes technological, logistic and human resources in wildfire fighting, assuring high performance and maximum flexibility thanks to its modular architecture, which is based on independent operative modules and on embedded communication system. Each sensor is equipped with a computational module that is responsible for data acquisition and metadata integration. Acquired data are radiometric images, jpeg images, video streams, geographical metadata and weather metadata. Data and metadata are transferred to the central server, which runs the image interpretation (fire risk evaluation, hot spot detection and smoke detection), the data storage, the alarms management and the decision support product generation. The user terminals access the central server for the evaluation of the statistical analysis of data stored in the database, for manual control and survey operations and for calibration and diagnostic operations. After the image interpretation, the central server sends confirmed alarms to the responsible agencies and operators as SMS messages, e-mail messages and the activation of signalling devices. The radiometric images represent the thermal distribution of the monitored scenario and are processed for hot spot detection. Jpeg images represent the chromatic distribution of the monitored scenario and are processed for the smoke detection, the false alarm reduction in hot spot identification and the visualization of the monitored scenario. Metadata are ancillary information concerning local time, geographical position, sensor orientation, weather data and system operative conditions.

4 FIRECAST®

FIRECAST® [1] is a computing system for forest-fire-danger-index forecasting, which elaborates weather parameter maps to evaluate fire danger indicators in the area of interest. FIRECAST® uses as a starting point the previsional Canadian Fire Weather Index (FWI), adjusted for continental Europe latitudes and climatology according to [2–7], and adapted for alpine region orography. The system improves the danger estimation by evaluating orographic parameters,

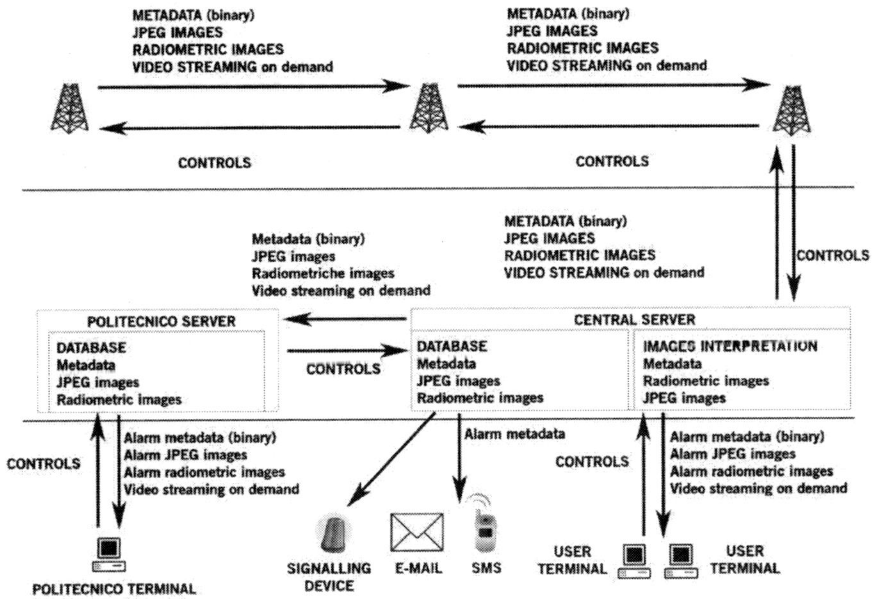

Figure 1: SIRIO architecture diagram.

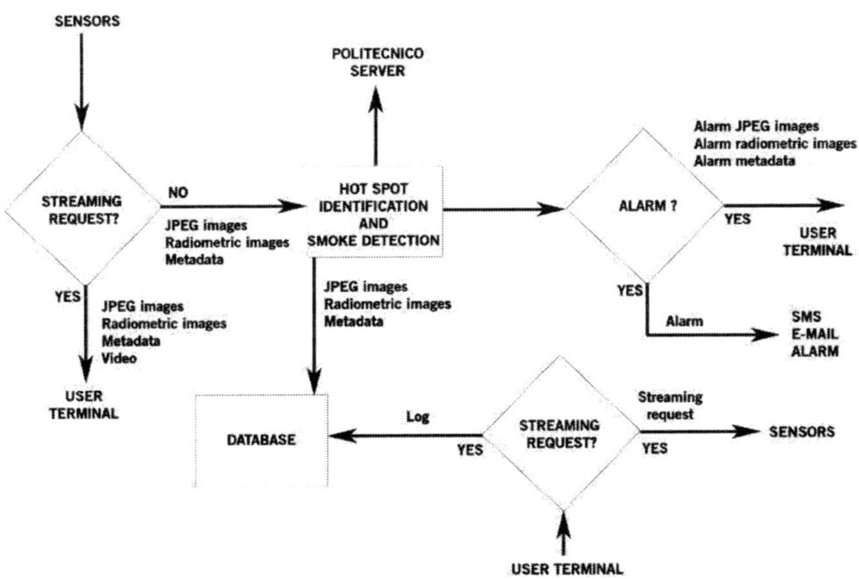

Figure 2: Server operational diagram.

such as terrain slope and orientation. Since the FWI is a meteorological index, it represents fire danger levels only due to present and past weather conditions, not considering contingently human presence and actions. FIRECAST® operates on meteorological forecast input data maps, in order to obtain output maps representing expected fire danger on the examined area with a forecasting time interval of up to 72 hours. To compute the final indices, the method also uses the historical evolution of these quantities. As explained in [8], the FWI system is composed of six codes, representing the daily changes in the moisture content of three classes of forest fuels with different drying rates: the rate of spread, the assumed fuel weight consumed and the fire intensity. In order to integrate input weather data with spatial variability information, FIRECAST® introduces correction factors related to slope (terrain inclination with respect to horizontal direction) and aspect (cardinal direction of surface's normal) in fire risk evaluation. Output fire risk is represented using four danger classes: EXTREME, HIGH, MODERATE and LOW. The validation results highlight the excellent capability of the system in forecasting reliable fire danger estimations and, most of all, in precise positioning of the alarm zones, with a good protection from false alarms. FIRECAST®, by evaluating combined fire risk simulation and fire statistics computation over the investigated region, allows the system to automatically select critical areas to be monitored. Furthermore, the previsional evaluation of the fire risk has an addressing function on surveillance and monitoring activities.

5 Hot spot detection system

The core of the hot spot identification algorithm is a radiometric model implementing a tailored progressive thresholds system that is applied to a combination of different frequency band (VIS, NIR, TIR) images related to the same scenario. The radiometric model evaluates the sensed scenarios by the integration of radiometric, climatologic, environmental, meteorological, orographic and vegetative characters with the sensor technical specifications. The model is based on a DEM and allows the tailoring of the identification method on the territory to be monitored. The model settings are customizable by the user, who can programme the territory analysis on the basis of specific monitoring requirements and of particular characters of the area to be monitored. The model is set when the system is installed and it operates during the monitoring activities in order to be updated to the current conditions. The radiometric model evaluates the sets of multispectral images (related to the same scenario) to be processed and automatically defines the best set of identification thresholds on the basis of the overall conditions. This procedure is applied to any set of sensed images. The system is thus capable of the identification and localization of fire hot-spot pixels. Often, particular 'non-dangerous' features and elements (e.g. sky, houses, farms, bridges, rivers, etc) occur in the sensors' field of view. The radiometric behaviour of these elements could affect the performance of the identification system. The algorithm features a masking tool in order to eliminate the 'non-dangerous' elements from the images, thus

Table 1: Hot spot detection performance.

HOT SPOT IDENTIFICATION			
KIA	POD	POFD	FAR
0.58	0.74	0.15	0.30

reducing the false alarm rate and the computational load of the automatic procedure. The masking tool is completely programmable by the user, and masks can be added or removed in a moment. Table 1 shows the identification performances of the module (KIA: K Index of Agreement – POD: Probability of Detection – POFD: Probability of False Detection – FAR: False Alarm Rate).

6 Smoke detection system

The presence of smoke, and therefore its early detection, is crucial as it is the first reminder or warning that an outbreak is about to degenerate. In many cases, the flame may not be easily seen and not detected by a hot-spot detection algorithm, as for example the burning of underbrush. In order to achieve low false alarm and missed detection rates, the Hot Spot Detection System outputs are processed by the Smoke Detection System, which evaluates images in the VIS and/or NIR domains. The smoke detection algorithm examines chromaticity changes and spatial and temporal patterns that characterize the smoke dynamics at an early stage of development. In order to detect the sudden irruption of smoke in the images, the system performs a two-step analysis on images. According to [9], the Blue (B) component of the RGB matrix has greater sensitivity to the changes generated by smoke in areas in which vegetation is predominant. The *static* block detects sudden increases in the B component with respect to a reference image. The *dynamic* block processes the images labelled with one or more alarm pixels by the static block output and through spatial and temporal correlations isolates effective smoke plumes from other moving features (birds, airplanes, etc.), thus reducing false alarms that may occur at the first stage of the process. At time t=N*Tc, where Tc is the image's sample interval, a set of N images is processed by the static and the dynamic blocks in order to reduce the false alarm rate. This phase is called the *detection phase.* At the end of this phase, the *confirmation phase* starts and static and dynamic blocks process a set of M images in order to eliminate any remaining false alarms. At the end of this phase (N+M images) an alarm is sent if smoke dynamics appear on the scene. After the initial warm up phase, the system is able to send an alarm every N-M images. Every Tr, a new reference images is loaded to prevent errors that could occur due to illumination changes throughout the day. Algorithm tests show very high reliability and robustness in the detection process. Combined with the hot spot detection, the smoke detection system enhances the fire rising detection and early warning efficiency, as depicted in Table 2 (KIA: K Index of Agreement – POD: Probability of Detection – POFD: Probability of False Detection – FAR: False Alarm Rate).

Table 2: Combined smoke and hot spot detection.

SMOKE DETECTION				COMBINED SMOKE AND HOT SPOT DETECTION			
KIA	POD	POFD	FAR	KIA	POD	POFD	FAR
0.77	0.89	0.12	0.16	0.84	0.92	0.12	0.06

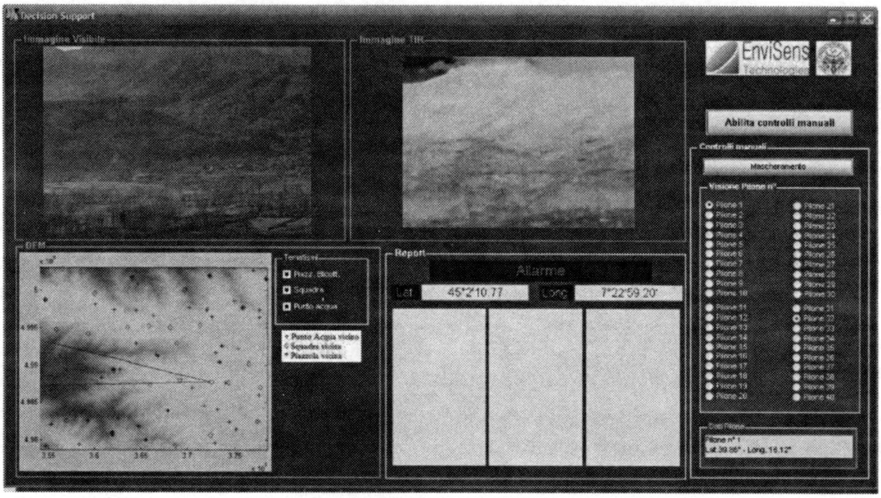

Figure 3: Decision support system user interface.

7 Decision support system

Forest fire prevention, monitoring and extinguishing operations in complex orography regions present dramatic problems related to hard environmental conditions, low population density and accessibility limitations affecting intervention activities and damage assessment. SIRIO operates an innovative projective geo-referencing algorithm that is able to geo-reference complex orography regions. As the Decision Support System is based on the evaluation of DEMs, it does not need the collection of Ground Control Points, which is a very hard task in complex orography environments. The algorithm is built up of three modules: sensor lens aberration correction, field of view localization on DEM and image geo-referencing. At the output of the system, each image pixel is linked to its Lat/Lon and UTM coordinates. The final products are a collection of scenario images sensed in the visible band, geo-referenced images with highlighted alarm pixels with overlaying DEM levels and topographic layers containing information to be used in the case of intervention. The final images contain geo-referenced information about sensors' field of view on the analysed area, hot-spot positions, helicopter landing spot positions, water supply positions, intervention squad localization, roads and so on. When fire is detected, the system identifies fire latitude and longitude, indicates accessibility to hot spots and puts in evidence territory characteristics and available resources. According to [10], the lens aberration correction algorithm evaluates and compensates for the sensors lens aberrations: astigmatism, curvature of the field, spherical aberration, geometric distortion and chromatic aberration. The system localizes the sensors' field of view on the DEM and applies the geo-referencing algorithm on the sensed images. The procedure is based on projective and geometric methods in order to achieve the best geographical linking trade-off. Elevation profiles and geographical information are extracted from the cone of view on the DEM. The decision support system presents a very friendly interface that allows for easy programming and makes management and intervention plans effective.

8 Conclusions and outlook

The SIRIO integrated system has been tested over different monitoring sessions and test areas and it is now operative in Piedmont and Liguria for wildfire monitoring and early warning. SIRIO achieves a high standard of performance in reliability, robustness, flexibility, cost and consumption. It guarantees accurate hot spot and smoke identification and produces geo-referenced information sets that are very useful for effective decision support activities. The RSG is under continuous development; at present, a new false alarm reduction tool based on combined VIS/NIR images is in the process of being implemented.

Acknowledgements

The present work has been produced as result of a wide research activity promoted by the Remote Sensing Group of Politecnicodi Torino, Regione Piemonte and Istituto Superiore Mario Boella. The research has been developed with the collaboration of EnviSens Technologies S.r.l. (www.envisens.com) and SVM S.r.l. (www.svm.it). The VM95® controller has been developed and implemented by SVM S.r.l., FIRECAST® and MicroRadarNetwork MRN® has been developed and implemented by EnviSens Technologies S.r.l.

References

[1] Corgnati L., Gabella M., Perona G., FIREcast system - Previsional fire danger index computation system for alpine regions. *1st International Conference on Forest Fires 2008, 17-19 September 2008, Toledo (Spain).*

[2] Viney, N.R., Hatton, T., Modelling the effect of condensation on moisture content of forest litter, *Agricultural and Forest Meteorology* **51**, 1990.

[3] Simard, A.J., The moisture content of forest fuels. A review of the basic concepts, *Canadian Department of Forest and Rural Development, Forest Fire Research Institute, Ottawa, Ontario*, Information Report FF-X-14, 1968.

[4] Van Wagner, C.E., Pickett, T.L., Equations and FORTRAN program for the Canadian Forest Fire Weather Index System, *Canadian Forestry Service, Petewawa Forest Experimental Station, Chalk River, Ontario,* Technical Report 33, 1985.

[5] Viney, N.R., Catchpole, E.A., Estimating fuel moisture response times from field observations, *International Journal of Wildland Fire*, 1991.

[6] Rothermel, R.C., Wilson, R., Morris, G., Sackett, S., Modelling moisture content of fine dead wildland fuels: input to the BEHAVE fire prediction system., *United States Department of Agriculture, Forest Service, Intermountain Research Station Ogden, Utah*, Research Paper INT-359, 1986.

[7] Nelson, R.M, A method for describing equilibrium moisture content of forest fuels, *Canadian Journal of Forest Research* **14**, 1984.

[8] Van Wagner, C.E, Development and structure of the Canadian forest fire weather index system, *Canadian Forestry Service, Petewawa Forest Experimental Station, Chalk River, Ontario*, Technical Report 35, 1987.

[9] Fernández-Berni, J., Carmona-Galán, R., Carranza-González, L., A vision-based monitoring system for very early automatic detection of forest fires, *Institute of Microelectronics of Seville - Centro Nacional de Microelectrónica Consejo Superior de Investigaciones Científicas y Universidad de Sevilla*, 2008.

[10] Losso, A., Corgnati, L., Perona G., False alarm reduction in forest fires detection with low-cost commercial sensors, *Gi4DM, Torino, Italy 2-4 February 2010.*

Innovative image geo-referencing tool for decision support in wildfire fighting

A. Losso, L. Corgnati & G. Perona
Turin Polytechnic, Italy

Abstract

Forest fires prevention, monitoring and extinguishing operations in complex orography regions present dramatic problems related to hard environmental conditions, low population density and accessibility limitations affecting interventions activities and damage assessment. In the scientific landscape, there are many existing image geo-referencing tools. They are developed in order to rectify images, which come from fixed ground station and/or satellite platforms. Some of these use satellite images to geo-reference complex orography regions and ground images to geo-referencing plane orography regions by using Ground Control Point collections. The present work describes an innovative projective geo-referencing algorithm able to geo-reference complex orography regions using a fixed ground station images. Besides, it does not need the collection of Ground Control Points, which is a very hard task in complex orography environments. The system is an innovative image geo-referencing tool conceived for decision support aid for wildfire fighting in alpine regions. The system is operating within a monitoring system in the Piedmont area (Northwestern Italy) and operates on images sensed by a fixed monitoring network. The geo-referencing software operates with a geometric and projective algorithm based on a Digital Elevation Model (DEM). The algorithm is built up of three main modules: sensors lens aberration correction, field-of-view localization on DEM, image geo-referencing. As said before, the system does not need any collection of Ground Control Points to rectify images. At the output of the system, each image pixel is linked to its geographical (Lat/Lon and UTM) coordinates. The final products are: a collection of scenario images sensed in the visible band; geo-referenced images with highlighted alarm pixels with overlaying DEM levels; topographic layers containing information to be used in case of intervention. Final images contain geo-referenced information about the sensors' field of view on the analysed area, hot-spot positions, helicopter landing spot

WIT Transactions on Ecology and the Environment, Vol 137, © 2010 WIT Press
www.witpress.com, ISSN 1743-3541 (on-line)
doi:10.2495/FIVA100161

positions, water supply positions, intervention squad's localization and so on. When fire is detected, the system identifies fire latitude and longitude, indicates accessibility to a hot spot and puts in evidence territory characteristics and available resources. The algorithm is part of a more complex system in which it complements an integrated fire risk evaluation and monitoring system, in order to enhance early warning processes and intervention and to more efficiently manage damage assessment.

Keywords: forest fires, fire monitoring, rectification, geo-referencing, image processing, alpine regions.

1 Introduction

Forest fires fight is a critical issue to preserve our environmental heritage. It is necessary to develop systems able to enhance the intervention efficiency and the operators' safety in case of fires too.

The present work describes an innovative tool within a forest fires monitoring system located on complex orography areas where the fire risk is high [1]. Each monitoring station is equipped with commercial sensors able to acquire and process images sensed on several frequency bands as: visible, visible + Near Infrared (NIR) and Thermal Infrared (TIR).

The sensors are piloted by an engine, which permits to scan the area of interest.

The system is able to automatically generate wide shot format to be sent, through GPRS or satellite communications, to the central server station.

Geo-referencing tool assigns each pixel's image to its geographical coordinates using geometric and projective methods on DEM. In particular it is the final part of an integrated forest fires detection system. The integrated system is built up of several interpretation and detection algorithms. The algorithms identify an alarm on images in which pixels contain an event of fire as flame or smoke [2, 3]. Therefore, in case of alarm, the fire detection output is processed by geo-referencing method which identifies the exact fire latitude and longitude. Moreover, topographic layers contains DEM's additional information as: camera field of view, helicopter landing spot's positions, water supply positions,

Figure 1: Monitoring station.

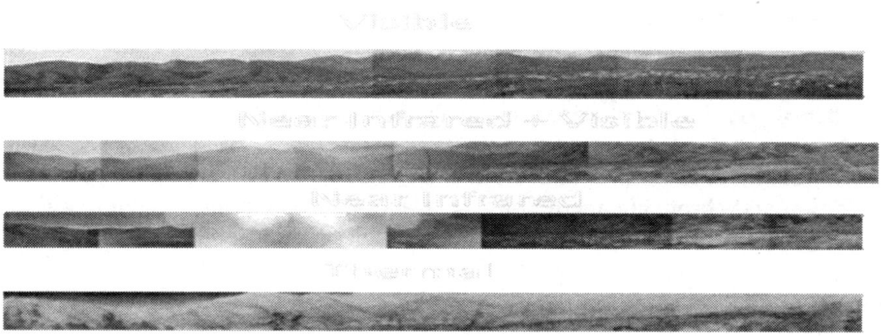

Figure 2: Example of frequency band sensed by monitoring station.

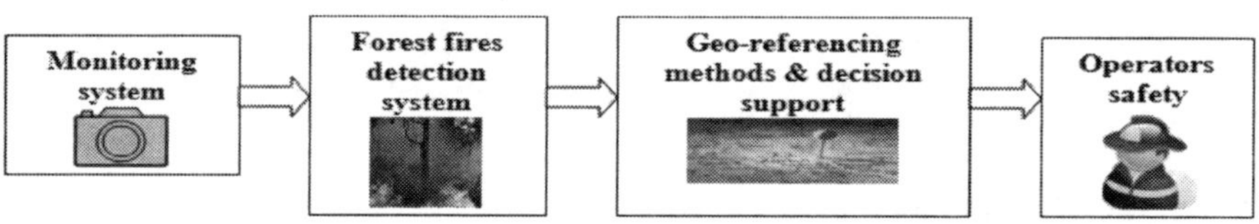

Figure 3: General block diagram of geo-referencing tool.

interventions squad's localization, service roads. If an alarm occurs, a complete set of geographical and logistic information is sent to the competent agencies.

Concerning early warning, the geographical position and the additional information are crucial for the early fire-extinguishing operations. In this paper we propose an innovative geo-referencing tool, which is able to generate geo-referenced imagery created from standard images collected by low-cost commercial.

The main steps of the procedure are the following:

- Derivation of camera's aberrations correction. The aberrations are errors introduced by the camera lens.
- Digital Elevation Model. The method localizes the camera field of view on DEM.
- Geo-referencing tool. The method is able to link image's pixel to geographical UTM coordinates.
- Decision support. The method generates additional information layers related to the event of fire.

The software is implemented in Matlab© and provides a graphical user interface. Results point out a very high reliability and robustness in the decision support process. The geo-referencing system is a valuable aid in fire fight management: it allows involved agencies for an efficient resources handling (both logistic and human), for a finalized territory monitoring and for an intervention planning oriented to operators' safety.

2 Lens distortion

Any photographic sensors, in particular commercial, could introduce a certain number of non-linear distortions due to aberrations in real lenses.

As a matter of fact, most frequent image imperfections (usually called aberrations) are: astigmatism, curvature of field, spherical aberration, geometric distortion and chromatic aberrations. In this work we take into account only the ones significant in our areas of work: the geometric distortion [4].

2.1 Geometric distortion

Geometric distortion is an aberration related to the position of points in the resulting image obtained from a real camera. It is crucial to consider distortion and aberration because they reduce the accuracy of measuring distances on the geo-referencing image so that they generate errors in the pixel position on image resulting in a displacement of image points with respect to their real positions.

Lens distortion is characterized by two main components called: Radial Distortion and Tangential Distortion.

2.1.1 Radial distortion

The geometric distortion affects the position of image points in the image. Using the distortion-free collinearity and some non-linear residual is possible to correct the errors caused by radial distortion. Two different types of distortion are possible, Barrel and Pin-cushion distortion.

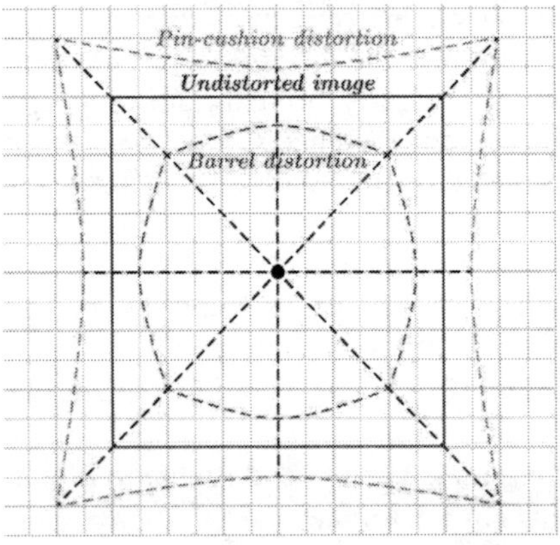

Figure 4: Barrel and pin-cushion distortion.

Basically, the barrel distortion causes the image points to crowd increasingly over the border and at the same time it causes the points close to the centre to spread on a radial direction.

The pin-cushion distortion presents opposite behaviour. It causes the image points to spread away when approaching the image border and at the same time it causes the image points to crowd when close to the image centre.

In order to correct geometric distortion a camera calibration tool is used.

Figure 5 (b) (c) (d) shows the pixel behaviour after aberration corrections. In particular the arrows show how the lens calibration operates on the external pixel, according to Barrel and Pin-cushion distortions.

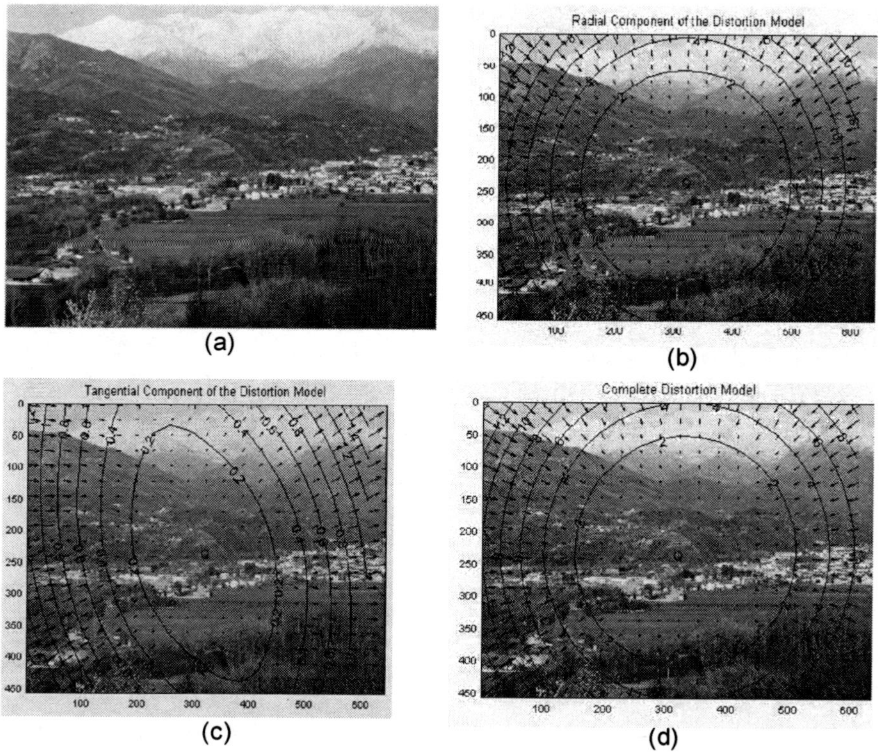

Figure 5: (a) Image affected by aberrations; (b) Radial correction; (c) Tangential correction; (d) Radial and Tangential correction.

3 Digital elevation model (DEM) and field of view (FOV)

The geo-referencing method does not use Ground Control Point collections to linked pixel in the geographical coordinates. In order to achieve our purpose we have to consider additional information and instruments. Concerning that, Digital Elevation Model is taken into account.

DEM is considered as a digital format territorial elevation distribution in which every single pixel is geo-referenced; in particular it is built up of raster format linking each pixel to its territorial elevation. Moreover, it is suited for Matlab© applications. In our case study the region to be considered is Piedmont areas.

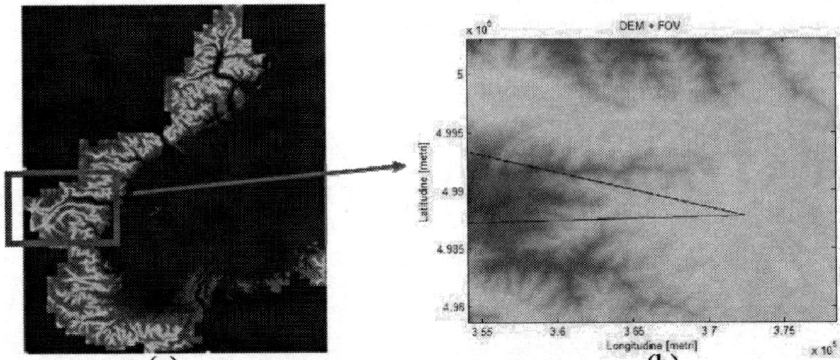

(a) (b)

Figure 6: (a) DEM (b) Field of view on portion's DEM considered.

Digital elevation model is useful to build up the camera field of view. According to this point small area is take in consideration. The field of view is represented as a cone, whose vertex is the monitoring system location. For this purpose, view angles, tilt, pan, orientation with respect to North and horizon elevations are crucial components for the field of view construction [5].

4 Geo-referencing tool and support decision

Once image aberrations correction and camera sensor cone of view are set on DEM, the geo-referencing method starts up. It is based on projective and geometric methods in order to achieve the best geographical linking trade-off [6, 7].

Elevation profile and geographical information are extracted from the cone of view on DEM; they represent the digital format view of images sensed by sensor. At the beginning, we just consider the horizon field of view and subdivide it into portions equals to horizon image resolution, as shown in Figure 7. For each portion the elevation profile is extracted, as shown Figure 8(a), the vertical field of view is subdivided into portion equals to vertical image resolution by using projective and geometric procedure, as shown in Figure 8(b). The selected portion is contained between two dotted lines, as shown in Figure 8(c).

Every single dotted line scans the vertical field of view from the bottom to the top a number of times equal to vertical image resolution. As it crosses the profile at least in one point, for each point it is possible to extract the relative geographical location, in particular the longitude location, fig 9. Basically, every line represents a certain vertical resolution value, linked to the longitude value. For example considering the first horizon and vertical scanning, if the first dotted line crosses the profile, the longitude value will be associated to the bottom left pixel's image and so on.

In order to find latitude component of the pixel just considered, we back up to the horizon field of view on DEM, fig 7(b).

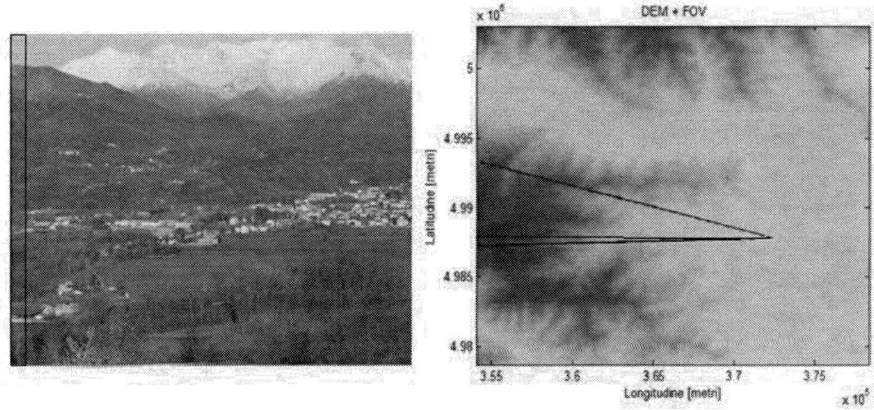

(a) (b)

Figure 7: (a) Example of field of view portion considered on image; (b) Example of horizon field of view portion considered.

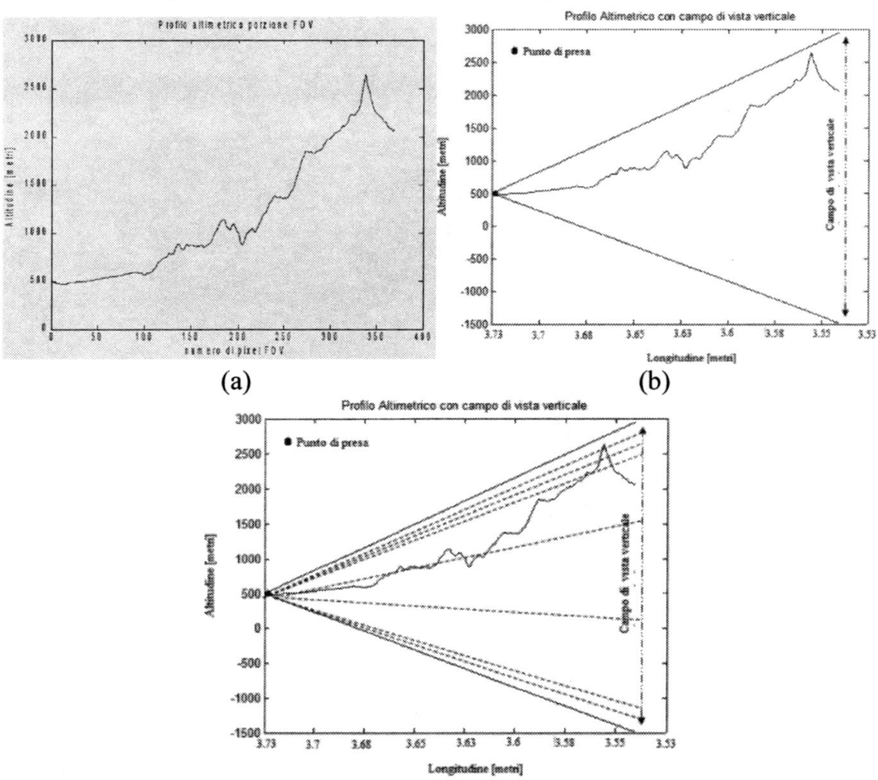

(a) (b)

(c)

Figure 8: (a) Portion field of view profile; (b) Vertical camera field of view; (c) Dotted lines equals to vertical resolution.

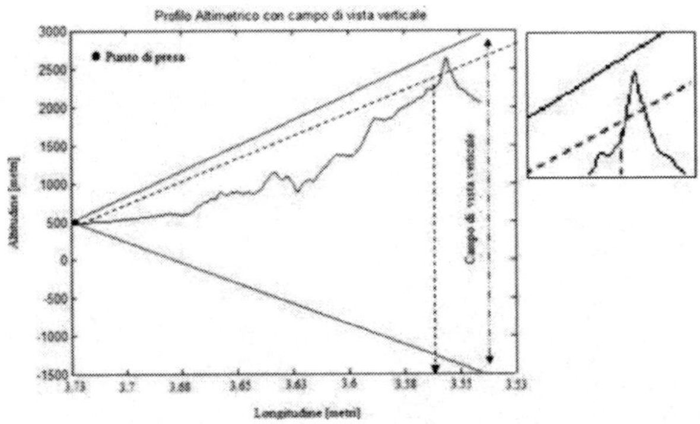

Figure 9: Dotted lines crosses the profile, determines longitude information.

At this point, knowing the longitude information, it is possible to determine the latitude information as shown on fig 10.

We trace a dotted line form to the longitude value until it crosses horizon field of view portion. The point determines the latitude value associated to the bottom left pixel, as mentioned in example.

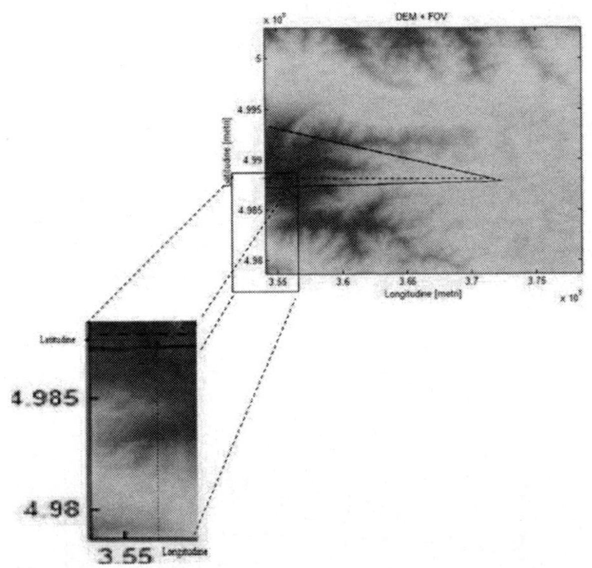

Figure 10: Determining the latitude component.

4.1 Decision support

In the case of a fire event, the forest fires detection algorithm identifies the pattern on the monitored scene. Geo-referencing tool is able to determine the exact geographical position of the event. At this point, it is possible to integrate additional information levels to be overlapped on DEM. The information levels are useful because they give additional tools to be used by operators. As an example, knowing the exactly geographical fire position it is possible to overlap on DEM logistic information, roads, helicopter landing, water supply positions, intervention squads, and so on. Moreover, all of them are geo-referenced and it is possible to choose them depending on the fire proximity.

5 Preliminary test and results

In order to test the algorithm, we built specific validation datasets made up of images taken from our test areas, located on Piedmont region. The main idea is to identify and mark some fixed points (FP) of known geographical positions within the images. For every single point, the real geographical coordinates have been checked using a GPS receiver in order to establish the mismatching between the real geographical coordinates and the results. As mentioned above, considering the image on fig. 5, we mark pixel coordinates, fig. 11, and the relative latitude and longitude are acquired by GPS receiver in the UTM geographical domain, tab. 1. For the best performance trade-off the image on fig. 5, has been converted in gray scale and masked on the areas where the fire risk is not significant, e.g. mountain snow cover.

The mismatches between real FP and geo-referencing method results are shown on fig. 12. The graph shows us the FP punctual difference and the avarage differences in latitude, longitude and distance in meters.

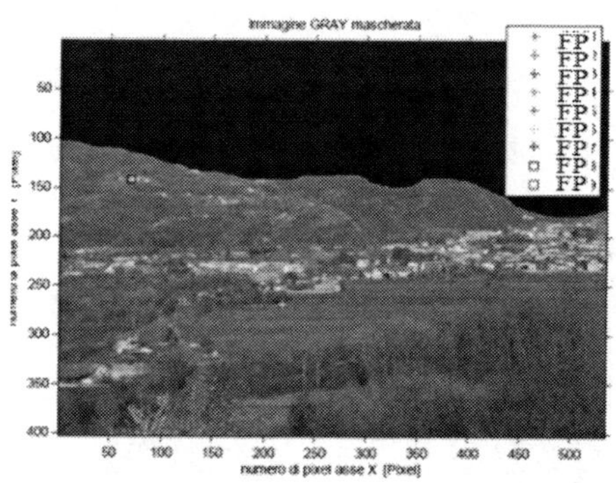

Figure 11: Image converted in grey scale, masked and marked with FP.

Table 1: FP pixel coordinates and the related true geographical coordinates.

	FP1	FP2	FP3	FP4	FP5	FP6	FP7	FP8	FP9
Pixel coord. X Y	7 330	531 232	321 238	293 283	384 245	218 268	6 221	70 142	528 247
UTM Latitude	49,877	49,882	49,881	49,879	49,88	49,878	49,876	49,876	49,882
UTM Longitude	3,717	3,706	3,708	3,712	3,706	3,712	3,691	3,678	3,709

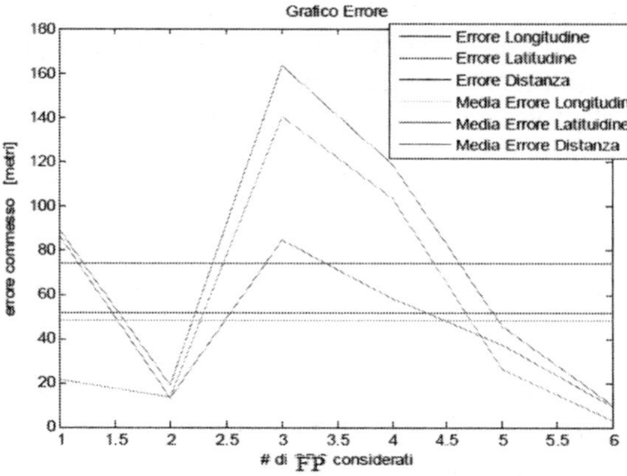

Figure 12: Mismatches in meters between true geographical coordinates and algorithm performance: Longitude (utm), Latitude (utm), distance, and average related them.

Considering decision support, we simulate an event of fire on the monitored scenarios. The algorithm carried out a report, which points out additional support information regarding helicopter landing, water supply positions, intervention squads closer to the fire, fig. 13.

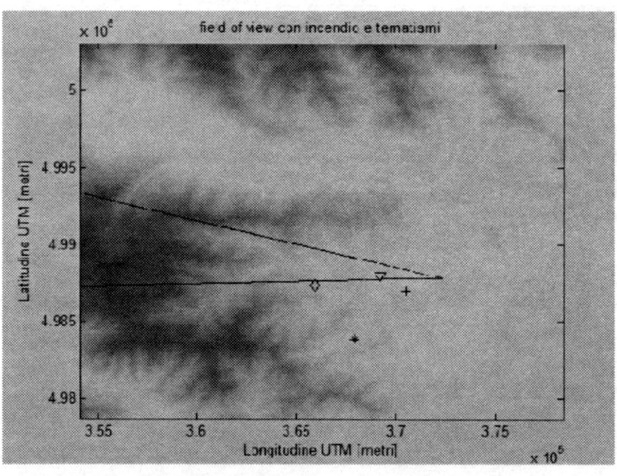

= Event of fire
= Intervention Squad
= Water supply
= helicopter landing

Figure 13: Decision support report on DEM.

6 Conclusion and future work

An innovative system based on automatic image geo-referencing methods for forest fire fighting has been implemented. Combining geometric analyses and projective transformations the system can link every single pixel to geographical UTM coordinates within time interval of few seconds with high spatial precision.

In order to validate the feasibility of the method, a set of images has been analyzed in our test areas. At the moment, our primary target is the successful implementation of the geo-referencing tool using images taken in the visible domain. According to the results shown above, the mismatches between true coordinate and algorithm results is on the order of less hundred meters. In particular, the average longitude error is 50 meters and latitude is more or less 80 meters, on the basis of the spatial resolution of the used DEM. Nevertheless spatial location mismatches are present due to several factors as: DEM and image resolutions, environmental conditions, device sensitivity, fire detection locations and so on. Future target will be spatial error reduction considering larger resolution size of both DEM and images. The algorithm is operating for test activities onboard the integrated monitoring system for high performance decision support SIRIO (Sistema Integrato per il Rilevamento di Incendi bOschivi) developed by EST S.r.l, Remote Sensing Group of Politecnico di Torino and SVM S.r.l.

References

[1] CORGNATI L.P.; GABELLA M.; PERONA G., FIREcast system - *Previsional fire danger index computation system for alpine regions, In: Modelling, Monitoring and Management of Forest Fires*, 1st International Conference on Forest Fires 2008, Toledo, Spain 17-19 September 2008

[2] Losso A. Corgnati L. Perona G., *early forest fires detection: smoke identification through innovative image processing using commercial sensor,* Environmental Including Global Change, Palermo, Italy 4-9 Ottobre 2009.

[3] Losso A. Corgnati L. Perona G., *false alarm reduction in forest fires detection with low-cost commercial sensors,* Gi4DM, Torino, Italy 2-4 Febbraio 2010.

[4] Nurc PO 4050852. Forward Eyes Video Image Rectification and Merging, 31 Dicembre 2005

[5] http://www.photorevolt.com/articoli_id_33.html

[6] Archetti R., Torricelli E., Erdman R., Lamberti A., *first application of a new imaging system for the coastal monitoring,* Bologna

[7] R. Archetti e A. Lamberti., *studio dell'evoluzione di una spiaggia protetta da opera a cresta bassa mediante videomonitoraggio*

Section 5
Resources optimization

Allocation of initial attack resources

D. B. Rideout, Y. Wei & A. Kirsch
Colorado State University, USA

Abstract

Increased scrutiny of federally funded programs combined with changes in fire management reflects a demand for new fire program analysis tools. We formulated an integer linear programming (ILP) model for initial attack resource allocation that operates in a performance-based, cost-effectiveness analysis (CEA) environment. The model optimizes the deployment of initial attack resources for a user-defined set of fires that a manager would like to be prepared for across alternative budget levels. The model also incorporates fire spread, multiple ignitions, simultaneous ignitions and monitoring resources on a landscape. It also evaluates the cost effectiveness of alternate fire fighting resources and alternative pre-positioning locations. Fires that escape initial attack are costly during the extended attack phase of fire management. To address this within the scope of initial attack, we constructed and analyzed alternative objective functions that incorporate a proxy for internalizing the cost of fires that escape initial attack. This type of model can provide the basis for a wider scale formulation with the potential to measure an organization's performance and promote a higher level of accountability and efficiency in fire programs.

Keywords: integer programming, initial attack, wildland fire, optimal deployment.

1 Introduction

Wildland fire organizations, including US federal land management agencies, customarily organize the suppression of unwanted fires into the three stages of suppression: initial attack (IA), extended attack (EA) and large fire management. Compartmentalizing this problem allows organizations to focus on the functioning and funding of different stages of fire management. This enables the analyst to focus with depth on the part of the problem of primary interest, but it introduces the problem of potential "spill over" effects that can be costly. For

WIT Transactions on Ecology and the Environment, Vol 137, © 2010 WIT Press
www.witpress.com, ISSN 1743-3541 (on-line)
doi:10.2495/FIVA100171

example, initial attack fires that are not contained will spill over into extended attack or even to large fire management. The potential cost of such spill-over is a necessary consideration in a proper benefit and cost calculus of initial attack.

We build on previous optimization literature to address the issues of multiple fires, simultaneous fires, monitoring resources, with special attention to potential spillover effects from IA to EA. A demonstrative example shows how an ILP model can be used to identify and optimize the dispatch of initial response resources in a performance-based and cost-effectiveness analysis (CEA) framework. The analysis includes four important features that have not been previously demonstrated: 1) use of an integer linear program (ILP) to model a functional relationship between cost and performance, 2) inclusion of multiple fires and optimal dispatch locations, with the potential to address a season of fires 3) the capability of including simultaneous ignitions, and 4) because fires that escape initial attack can be costly, we address alternative means of including a proxy for the cost of fires that escape initial attack. The remainder of the paper is structured as follows: in the next section we present a description and a mathematical formulation of the ILP with alternative objective functions to address the cost of escaped fires, this is followed by a demonstrative numerical example to illustrate the capabilities and relationships of the model. The last section provides discussion and conclusions including model limitations and potential extensions of the formulation.

2 A performance-based fire preparedness ILP

We make the customary assertion of minimizing damage for a given level of expenditure consistent with the least cost plus loss expressions (Rideout and Omi [3]). Consistent with this assertion, we compare the effectiveness of alternative initial attack organizations by minimizing expected damage (loss) of unwanted wildland fires for any specified budget level where a range of budget levels are modelled. We recognize that to the extent that firefighting resources are scarce, not all fires are of equal importance to contain because not all resources that could be damaged by fire are of equal consequence. Wildland fires that occur in the wildland urban interface threaten life and property are typically of greater importance to aggressively manage than are fires occurring in remote areas such as wilderness. Because acres differ in their importance to protect from wildfire, our formulation provides the ability to proportionally weight acres that might be differentially affected by the damaging effects of wildfire (Rideout et al. [4]). The calculation of natural resource loss for a given budget level involves multiplying the area burned from each fire by its per acre weight to calculate the per acre loss. The weight reflects the marginal rate of substitution of resource disimprovement. The ILP optimization allows us to focus on cost effective solutions while avoiding interior (inferior) solutions.

A set of fires is provided as input to the ILP and each fire includes information on its initial size and its change in perimeter and area by time period. Perimeter is directly related to suppression cost through resource production rates and the area burned is directly related to performance through expected loss. We use the free burning fire containment rule from previous deployment models (for example, NWCG [5]) stating that a fire is contained when the total fireline produced by firefighting resources overtakes the fire perimeter. A fire is defined as having "escaped" if it is not contained during the initial attack period due to a lack of funds to apply to fire fighting resources resulting in a lack of sufficient fireline production capability.

A pool of potential firefighting resources is established for evaluation where each resource can be selected and optimally allocated to a set of candidate dispatch locations and fire events. Each firefighting resource is defined by a fireline production rate and by its fixed and variable costs. Fireline production is modelled by a cumulative value that is input for each time step of each fire. An advantage of the discrete time step approach is that the production function does not have to be constant or linear. Thus, production rates can reflect fatigue and other disruptions in production such as water and fuel refills. Arrival times and travel delays can also be reflected in these production values by entering zero chains of fireline production during travel periods. The model uses the production information along with other factors to solve for the optimal deployment.

The costs of initial response resources and of fire escapes are important considerations in preparedness modelling that directly impact the preparedness budget. This ILP model inputs fixed and variable costs of firefighting resources that directly impact optimal deployment. The fixed cost is modelled as a one-time charge that is incurred if the resource is deployed to any fire during the season. Each resource's variable cost is modelled as an hourly cost that reflects its operating expenses on each fire including maintenance, fuel, regular hourly wages, overtime and hazard pay. Also during the IA period, we deploy a monitoring resource to escaped fires to reflect the concept that every fire, contained or not, will receive some monitoring efforts during initial attack. The full cost of escapes is addressed in the section "Incorporating a Proxy for the Cost for Escaped Fires".

Formulating the ILP for fire suppression requires developing a set of equations to track containment on each fire. The ILP optimized firefighting resource allocation to a single fire to minimize the total suppression cost plus net value change. They used a separate set of constraints at each time period to track whether the targeted fire would be contained during that period. This formulation expands their approach to support IA firefighting resource allocations across multiple and simultaneous ignitions. Although firefighting resources can be dispatched to multiple fires, they often cannot be dispatched to simultaneous fires, and this introduces heightened competition for firefighting resources. To model simultaneous ignitions, we forced each resource to choose one of the simultaneous ignitions to attack and we assumed that resources would not be redeployed to other simultaneous ignitions. This restriction reflects the pragmatic consideration that ground based resources often lack the mobility to address simultaneous fires. We also introduced constraints across time to track the time that each fire was contained.

$$Loss = \sum_{i} \sum_{d \in (1 \text{ to } D_e)} (W_{id} * f_{id} * A_{id}) \tag{1}$$

i index of a fire in the set I of all fires,
d index of fire durations in the set D and D_e. D is the set of periods before a fire escape. D_e is the period at which fire is considered to have escaped,
f_{id} binary variable, fid = 1 if fire (i) burns for a duration of (d) time periods, otherwise fid=0,
W_{id} predicted fire losses for each unit of area burned by fire (i) after a duration of (d) time periods,
A_{id} total area burned by fire (i) for the duration of (d) time periods,

The objective function (1) minimizes the expected fire loss for a given budget and each firefighting resource is restricted to a single location. This expands the model to consider alternative locations for any particular resource. Each suppression resource can only be deployed to each fire for a fixed duration and each fire lasts for a single duration. Additional restrictions are available from the authors. For each contained fire, the total length of fireline produced by all suppression resources from different dispatch points must equal or exceed the fire perimeter at the period it is contained. We also ensure that fireline will be effective only during the containment period of any fire. The index of fire duration d is used to make this assumption valid in the model. For example, if there is a single suppression resource r' available of constructing fire line to contain a fire within an 8-hour IA period, this constraint will take a simplified form of:

$$f_{r'1}+2f_{r'2}+ \ldots\ldots +8f_{r'8} >= x_{r'1}+2x_{r'2}+ \ldots +8x_{r'8} \tag{2}$$

2.1 Incorporating a proxy for the cost of escaped fires

While compartmentalizing suppression into IA and EA provides managerial clarity for planning, budgeting and operations, it introduces a classic externality problem if not properly addressed. In the IA preparedness planning context, such an externality can be generated if the costs of fires that escape IA are not considered in the IA model or decision process. A correct approach, consistent with the Coase Theorem (Coase [2]), would be to maximize the sum of the net benefits across both program components (IA and EA) when considering resource allocations to IA preparedness planning. Simultaneously modelling both would, in principle, provide the correct set of costs to the IA analysis. In this way we could solve for the optimal number of escaped fires. The problem is that there is no precedent for modelling large fires in this context or for modelling IA and EA simultaneously.

In lieu of a credible simultaneous solution, we tested three potentially practical proxies for the cost of escaped fires by using three alternative objective functions. These were: A) using a large per escaped fire penalty, B) increasing the penalty for escapes in proportion to estimated loss at the time of escape and

C) combining approaches one and two. The objective function is separated into two parts where the first part represents the loss during IA and the second part represents a penalty for escapes. Of particular interest is how modifying the second part of the objective function will influence the allocation of IA resources and fire containment.

Objective function A) penalizes each escaped fire by using a large constant penalty "M". As M becomes large, this objective function effectively maximizes the number of fires contained, regardless of their importance. This is also known as initial attack success rate: a common performance metric. In objective function B), escaped fires are penalized by a value proportional to their loss right before escape.

The penalty increases linearly with respect to loss and the term $K \geq 1$ enables us to increase the magnitude of the penalty. The rationale for penalizing escapes based upon the estimated loss at the time of escape is that it reflects the last information known to the IA model regarding the potential resource damage from an escape. It also reflects the restriction of the scope of the problem to IA preparedness.

Objective function C) combines A) and B) to penalize escapes by using a constant penalty combined with the estimated loss prior to escape. The rationale for adding the per fire escape cost is that escaped fires can be costly to manage even if there is little potential for resource loss at the time of escape.

With the loss minimizing ILP formulated and expressed through three alternative objective functions to address the cost of escapes, we apply the model to a demonstrative example that is designed to show how the model addresses optimal placement and dispatch of resources in a CEA context at different budget levels.

2.2 Demonstrative example

We begin by defining a fire scenario that includes 10 fires where two occur simultaneously. For simultaneous ignitions we make the simplifying assumption that no single suppression resources can be assigned to both. This assumption can be relaxed to allow some resources to serve simultaneous fires, but such relaxation does not add to the substance of our findings or formulation. We also assume eight time periods where each period is one hour. The duration can take any time step and the time steps are not required to be uniform. The initial perimeter of each fire represents the size of each fire when at discovery and the perimeter of each fire will grow as defined by the user during the eight hour IA period.

Our list of fire fighting resources was selected to illustrate key model features of optimal allocation and dispatch while recognizing that agency planning units would be considerably more complex. For demonstration we model three kinds of resources: resources that are relatively inexpensive and have relatively low production rates such as handcrews, resources that are moderately expensive but produce greater line production such as engines, and we also included dozers as an expensive and highly productive resource. Resource production rates were based on the National Wildfire Coordination Group Fireline Handbook

(NWCG [5]). To demonstrate the model's ability to evaluate optimal resource placement, we allowed the model to choose from three possibilities: dispatch from location HC1.A, dispatch from location HC1.B, and no dispatch. The difference in dispatch locations is represented by differences in arrival times and by the subsequent fireline production on each fire. The cost and productivity of each kind of resource was used. By using the firefighting data on fire growth, expected loss and firefighting production, we generated the following results.

3 Results

The results of the model formulation using the demonstrative example are discussed in two parts: 3.1) model formulation on resource allocation and fire containment with the effects of simultaneous fire events including the use of monitoring resources, and 3.2) effects of the alternative objective functions reflecting different proxies for the cost of escaped fires.

3.1 Resources and fires

The detailed containment period for each fire and the allocation and dispatch schedule for each resource are based on a budget level of $21M. At this budget level all fires can be contained and there was no difference among the alternative objective functions for escaped fire cost. In addition, the deployment duration of any resource is less than or equal to the duration of each corresponding fire. The necessary and sufficient condition of containing fire i at period d is that the total length of fireline produced for fire i at or before period d has to be equal to or longer than the perimeter of fire i at period d.

All fires, except for the simultaneous fires were contained within either the first or the second hour. The key advantage to containing fires earlier is to reduce potential loss. Keeping fires small also means that less fireline is needed and this should not imply a lower suppression cost because minimizing fire size implies an intensive effort that could employ the most expensive and productive equipment and labour. The results also show that handcrew 1 would be allocated to dispatch point B at this budget level and that handcrews 2 and 3 and engine 3 were also dispatched. The expensive and technically superior dozer was not dispatched at this budget level.

The ILP was required to make "tough" choices in resource deployment on the simultaneous fires. Simultaneous fire F10 used all of the hand crew resources while simultaneous fire F9 relied entirely upon engine three. The opportunity cost of deploying all of the handcrews to F9, in terms of reduced effectiveness on F10, is apparent as it took longer to contain F9 (five time periods). The cost of deployment includes both the variable cost of deploying the resource plus the opportunity cost incurred by not allowing that resource to attack the competing simultaneous fire. Additional tests showed that after removing the assumption of simultaneity for fires F9 and F10, a 100% IA success rate was achieved at a lower budget level of $18M.

3.2 Alternative proxies for the cost of escaped fire

Results from the model at the 11 different budget levels were analyzed. Our tests of the 11 budget levels show that all 10 fires can be contained at budgets of \$21M or above within the 8-hour initial attack period. Fire containment schedules are insensitive to the choice of objective function above this budget level. Reducing the budget increased scarcity and the model allowed some fires to escape. The number of escapes was influenced by the budget level and the objective function. For a given budget level, which fires escaped was sensitive to how the proxy cost of escapes was modelled.

Analysis of objective function A) showed that the model would contain as many fires as possible. That is, a simple per fire proxy for the cost of escapes maximized initial attack success rate. This objective function will always maintain or increase the number of contained fires with increases in the budget. However, using this kind of objective function produces dispatch schedules with higher fire losses during IA. Because its constant penalty treats all fires with equal importance for containment, it fails to recognize the relative importance between fires.

Objective function B) penalized each escape proportionate to its loss at the time of escape. Given the IA scope of the analysis, this might reflect the best, albeit imperfect, information available to the model. Weighted size reflects the last known information from IA regarding values at risk, the size of the fire, and the likely cost of managing fire in an EA setting. Here, with a budget level that is insufficient to contain all the fires, containment decisions reflect the relative importance of fires at escape. Test results, with K =1, show that as the budget increased from \$15M to \$16M, the number of escaped fires increased from three to five while the loss decreased from 1,771 to 1,429. With a \$1,000 budget increase, the model shifted from containing a group of five less important fires to a group of three more important fires. This local result reflects the possibility of encountering the economically inferior fire (fires that would not be contained at higher budget levels). Globally, however, as the budget increases so will the number of contained fires. In B) the value of K can be increased in an attempt to reduce the number of escapes, but this is nearly always futile because increasing the value of K does not change the relative importance between escaped fires. Increasing K had no effect on containment decisions in our example.

Objective function C) combines the costs from objective functions A) (cost per fire) and B) (loss at escape). By using a large constant penalty M the model will contain as many fires as possible, thus maximizing initial attack success rate. If there are multiple ways of containing the same number of fires, the model will select the most important fires. Model results also show that by using this objective function, as the budget level increased from \$ 15M to \$16M the initial attack success rate did not decline. This suggests that using objective function B) suffers from the same problem as objective function A), where five less important fires were contained but three more important fires are allowed to escape.

4 Conclusion

The ILP model developed in this paper includes several innovations while demonstrating key economic principles of optimal initial attack. The ILP expanded on previous work to address the planning principles for a set of fires. It shows how scarce firefighting resources would be allocated to alternative fires to minimize loss at any given budget or appropriation level. By addressing the allocation of resources across a set of fires, we enabled the model to identify which fires to fight and how aggressively to fight them. In this way, the model also demonstrated how optimal dispatch locations can be scheduled and how different kinds of firefighting resources might be utilized. Altering firefighting resource scarcity through budget levels also demonstrates how optimal results and their locations are dependent upon the level of the budget. Increases in the available budget allow for greater loss reduction and usage of more effective resources, but changes in the available budget can affect optimal location decisions. The management of scarcity is particularly important when simultaneous fire events are considered. While all fires compete for scarce resources across a planning season, simultaneous fires compete more intensively by effectively precluding the simultaneous use of individual firefighting resources. Our example showed how two simultaneous fires were managed differently by different kinds of resources to minimize overall loss.

Optimal resource use for initial attack requires that key cost elements are included in the model. These include the cost of having firefighting resources available (fixed cost), deployment costs (variable costs) and the cost of fires escaping IA. Managing the cost of escapes within the initial attack scope is inherently problematic because, by definition, they are external to the scope of analysis. Therefore, they can pose the classic externality problem if not properly analyzed. Because expanding the scope of analysis to extended attack (and potentially beyond), is currently infeasible, we analyzed three alternative approaches to include a proxy for this cost. The first proxy effectively maximized initial attack success rate by including a large per fire cost where all fires escape costs were treated equally. This resulted in important fires escaping under the constrained budget while relatively unimportant fires were contained. The second proxy introduced a cost based upon the loss at the time of escape. While this approach distinguishes between important and unimportant fires, a local consequence is that fewer fires may be contained as the budget increases. The principle applied is intended to reflect the potential cost and especially cost differences of fires that would escape. The technology applied to make these cost estimates could be greatly expanded through predictive fire behaviour modelling and GIS mapping to generate a reasonable estimate of escaped fire cost. However, improving the technology does not alter the principles of in this demonstration. The third proxy includes both costs modelled simultaneously. Since the priority of this proxy is to maximize the IA success rate, it could also allow important fires to escape while containing fires of lesser importance.

While the ILP was intended to demonstrate managerial principles of optimal resource use in preparedness planning, especially in initial attack, it serves several other purposes. First, it is a useful demonstration of key economic elements of optimal resource allocation across a set of fires or initial attack. Such a model can also serve as a framework for thinking about how decisions can be made in ways that are consistent with principles economic efficiency. Secondly, an ILP model can be augmented or modified in many ways. For instance, instead of using a single fire scenario, as we did here, multiple scenarios could be used. Other enhancements could include a stochastic analysis of modelling the uncertainties in size and cost of escaped fires, and the variations existed in fire line productivities. Optimal deployment models, such as the approach illustrated here provide potentially useful insights for understanding and illustrating the efficient use of scarce resources. While optimization models have strengths and weaknesses, capitalizing on the strengths may be best realized by combining optimization with other complementary approaches such as simulation.

References

[1] Aneja, Y.P., Parlar, M., 1984. Optimal staffing of a forest fire fighting organization. Can. J. For. Res. 14, 589-594. 1984.

[2] Coase, R.H., The problem of social cost. J. of Law and Economics. 34, 1-44. 1960.

[3] Rideout, D. and P.N. Omi. Alternative expressions for the economics of fire management. Forest Science. 36(3):614-624. 1990.

[4] Rideout, D. B., P. S. Ziesler, R. Kling, J.B. Loomis and S. J. Botti. Estimating Rates of Substitution for Protecting Values at Risk for Initial Attack Planning and Budgeting. For. Pol. and Econ. 10(2): 205-219. 2008.

[5] National Wildfire Coordinating Group Fireline Handbook (NWCG). 2004. NWCG Fireline Handbook. 3. PMS 410-1. NFES 0065. National Wildfire Coordinating Group. Boise, Idaho.

Optimal timing of wildfire prevention education

D. T. Butry[1], J. P. Prestemon[2] & K. L. Abt[2]
[1]Building and Fire Research Laboratory, National Institute of Standards and Technology, USA
[2]Southern Research Station, US Forest Service, USA

Abstract

Public outreach and wildfire education activities have been shown to limit the number of unintentional human-caused ignitions (i.e., 'accidental' wildfires). Such activities include the airing of public service announcements, visiting with homeowners in at-risk areas, distributing informative brochures and flyers, hosting of public forums (with presentations), and facilitating community sponsored neighborhood hazard assessments. By limiting the number of ignitions, prevention entirely avoids costs and losses related to fire suppression (initial attack) and property damage. In this paper, we show that the benefits of wildfire prevention education activities carried out in the State of Florida, USA, far outweigh their costs. We also report how the return on wildfire prevention education investment in that State varies—i.e., the effectiveness of these programs varies—with many exogenous factors, including weather, season, and recent wildfire history and fuels management activities. To illustrate how this effectiveness variation could be exploited to increase returns to money spent on prevention, we explore the optimal timing of wildfire prevention activities. Optimal timing of wildfire prevention education spending is defined as the spending allocation over time that yields the lowest wildfire-induced cost plus net value change to society. We find that, for Florida, the optimal (monthly) timing of prevention activities can be forecasted by exploiting the relationships between prevention effectiveness and fire weather measures, which vary predictably within the year.

Keywords: fire economics, wildland-urban interface, hazard mitigation, wildfire prevention, wildfire education.

WIT Transactions on Ecology and the Environment, Vol 137, © 2010 WIT Press
www.witpress.com, ISSN 1743-3541 (on-line)
doi:10.2495/FIVA100181

1 Introduction

Wildfire managers and policy makers have a variety of means for reducing the long-run discounted sum of costs and losses from wildfire. These include preventing and suppressing wildfires, reducing fuels so that fires are less damaging and easier to suppress, and taking steps after a wildfire to minimize the overall impact to a community or ecosystem. Although research has statistically identified the effectiveness of suppression (Butry [1]) and fuel reductions at reducing fire extent and damages (Pollet and Omi [2], Mercer *et al.* [3]) and lowering the rate of occurrence of certain types of fires (Mercer and Prestemon [4], Prestemon and Butry [5]), up until very recently (Prestemon *et al.* [6], Butry *et al.* [7]) the refereed literature had been missing studies documenting the effectiveness of wildfire prevention efforts directed at human-caused wildfires. Governments and other entities devote significant resources to educating the public about the dangers of, and ways to reduce, unintentional firesetting, but the economic justifications for such interventions have been tenuous.

Wildfires are unintentionally ignited through a variety of mechanisms, including by escaped use fire (e.g., debris fire, brush-clearing fires), equipment malfunctions or sparking, escaped campfire, smoking, children-involved fire play, and vehicle crashes. Some of these wildfire starts can be avoided, and agencies have long used policies to prevent them, such as not permitting use fires or campfires when weather conditions are favorable for fire spread. Land management agencies have also undertaken significant programs that involve public education and organization of communities, and this includes encouragement of adoption of less risky technologies (e.g., use of spark arrestors). The rationale behind such programs is that preventing ignitions reduces expenditures required for fuels management and suppression, and because unintentional fires are often due to carelessness, information campaigns can raise awareness. Further, unintentionally ignited fires often occur in places where people and values at risk are in close proximity (Butry *et al.* [8]). Because people are often involved in the wildfire start, and because these fires occur in an intermix of high value property and local populations (e.g., Bradshaw [9]), such fires produce immediate peril to people and property close by.

In this paper, we extend the methodologies and models developed in Prestemon *et al.* [6] and Butry *et al.* [7] that were used to: (1) evaluate the benefit-cost performance of wildfire prevention effort in the state of Florida (Prestemon *et al.* [6]), and (2) determine the optimal mix of wildfire prevention effort and fuels management to yield the least cost-plus-loss of wildfire management (Butry *et al.* [7]). We extend that research by estimating the optimal timing of wildfire prevention activities such that it yields the least cost-plus-loss while holding fuels management activity constant.

2 Wildfire prevention effectiveness in Florida

The effectiveness of wildfire prevention can be measured both in terms of the number of wildfires prevented per unit of wildfire prevention applied and in

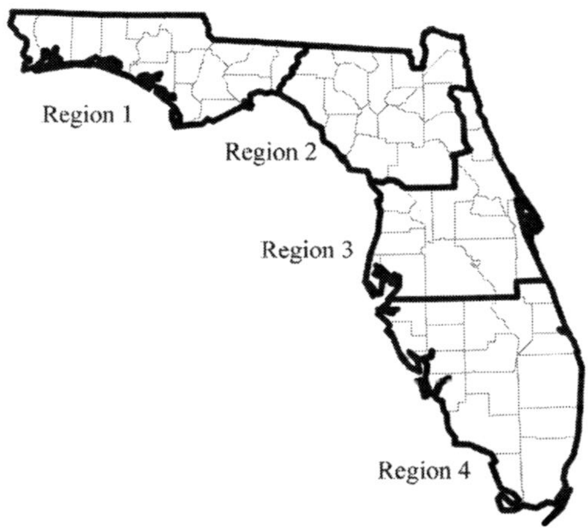

Figure 1: Fire management regions (black outline) and counties (gray outline) in Florida.

terms of damages averted per unit of wildfire prevention. Calculating the returns to wildfire prevention is straightforward and derives directly from equations of wildfire starts modeled empirically as a function of prevention and a variety of other variables, including fuels management, weather, historical wildfire, and socioeconomic measures.

Prestemon *et al.* [6] estimated the effect of wildfire prevention effort on the number of 'targeted' wildfire ignitions in the four wildfire management regions of Florida (see fig. 1) from 2002-2007.

Targeted wildfire ignitions include unintentionally ignited wildfires caused by escaped campfires and debris fires, discarded cigarettes, and children playing with fire. These wildfires accounted for 320 ignitions on average per year, over the study period, and occurred mostly during the winter (December, January, and February) and spring months (March, April, and May) (see table 1). Overall, targeted wildfires accounted for 36% of all wildfires in Florida during this time. Debris fires caused 23% of all ignitions, followed by campfires (7%), children (5%), and smoking (1%) (see fig. 2).

Five wildfire prevention activities carried out by wildfire mitigation specialists in the State of Florida were evaluated: (1) media public service announcements (PSAs) broadcast to the general public, which included the number of TV, radio, newspaper PSAs, and activities performed by wildfire prevention specialists, including (2) homes visited, (3) presentations given, (4) brochures distributed, and (5) community-based wildland hazard assessments. Although there were other kinds of prevention education activities recorded by wildfire mitigation specialists, these were too limited to enable their inclusion in the statistical analysis we conducted. The average timing of the five activities evaluated is shown in table 1. For all activities, except homes visited, their peak occurred in the spring months (March, April, and May). This peak coincided with the peak wildfire activity of all causes (targeted plus non-targeted wildfires). Note, however, many of the targeted types of ignitions preceded prevention and began in the prior winter. This suggests an opportunity for better timing of wildfire prevention activities, so that the message gets out before the relevant wildfire season begins.

Table 1: Annual wildfire prevention effort and fire activity in Florida, 2003-2007.

	Averages					
PREVENTION	***Spring***	***Summer***	***Fall***	***Winter***	***Monthly***	**Total**
Media PSAs	910	238	167	252	130	1,566
TV	400	89	53	89	53	632
Radio	180	72	54	71	31	377
Newspaper	330	77	59	91	46	557
Homes Visited	84	184	51	58	31	377
Presentations	13	8	7	18	4	45
Brochures	2,549	1,560	975	3,884	747	8,968
Assessments	2	2	2	2	1	8
FIRE						
Rx fire acres	47,025	12,349	10,492	69,900	11,647	139,766
Rx fire permits	298	54	60	473	74	886
Targeted Ignitions	143	27	32	118	27	320
Non-targeted Ignitions	232	172	47	95	46	546

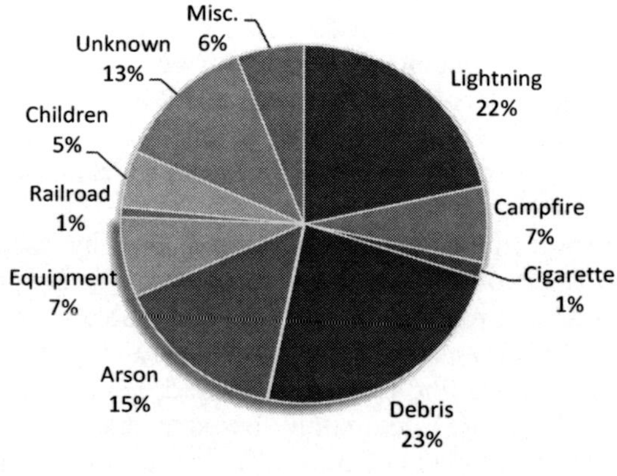

Figure 2: Percent of all wildfire ignitions by wildfire cause in Florida, 2002-2007.

3 Methods

Using the modeling results from Prestemon *et al.* [6] and Butry *et al.* [7], we explore how changes in the timing of prevention activities affect the social cost-plus-loss of wildfire management. These studies relate wildfire prevention effort to the number of targeted wildfire ignitions while accounting for other exogenous factors. Table 2 presents the elasticities of targeted wildfire ignitions with respect to wildfire prevention effort (i.e., the percent change in the number of targeted ignitions due to a corresponding 1% change in wildfire prevention effort) reported in Prestemon *et al.* [6]. This shows the effect prevention has on the numbers of fire starts. For instance, a 10% increase in media PSAs is expected to cause a 1.7% decrease in the number of targeted wildfire ignitions in the month that the PSAs were run, and another 2.6% decrease in ignitions over the following six months, for a total reduction of more than 13 fire starts. As a comparison, the elasticity of targeted wildfire ignitions with respect to prescribed fire ranged from -0.18 to -0.34, depending when treatments had last occurred (1 to 3 years prior). Thus, wildfire prevention and prescribed fire have similar elasticities, but because the effect of prescribed fire is longer lived, the total reduction is larger (over time, nearly 17 fewer fire starts for a 10% increase in prescribed fire).

Table 2: Elasticities of targeted wildfire ignitions with respect to wildfire prevention effort.

Wildfire Prevention Activity	Elasticity
Media PSAs: 1-6 months prior	-0.26
Homes Visited: 1-6 months prior	0.04 (not significant)
Presentations: 1-6 months prior	-0.22
Brochures: 1-6 months prior	-0.24
Assessments: 1-6 months prior	0.07 (not significant)
Media PSAs: current month	-0.17
Homes Visited: current month	-0.03
Presentations: current month	-0.23
Brochures: current month	-0.14
Assessments: current month	-0.12

Butry *et al.* [7] determined that a strategic coordination of wildfire prevention with prescribed fire treatments could be used to reduce the numbers of targeted wildfires in an economically efficient manner. Figure 3 shows the cost-plus-loss surface of wildfire management, drawn as a function of wildfire prevention effort and prescribed fire activity. It was found that increasing wildfire prevention by 168%, combined with increasing prescribed fire treatment by 74%, reduced the cost-plus-loss of wildfire management from $325 million to $301 million. This expansion of these wildfire management programs resulted in $24 million in net benefits to the state of Florida, and that no other combination of prevention and fuels management would deliver a larger economic return.

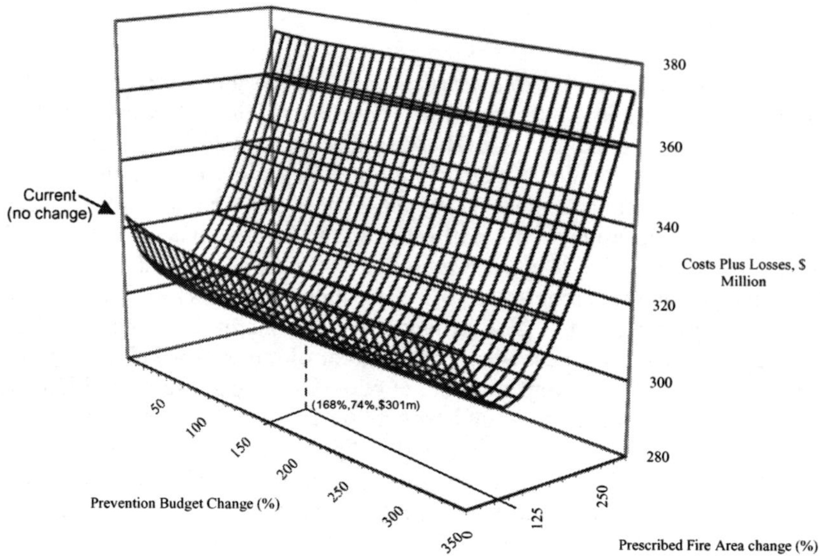

Figure 3: Wildfire mitigation trade-off: wildfire prevention effort versus prescribed fire fuel treatments.

Of course, actually expanding the number or size of prescribed fire treatments in Florida may be difficult. Unlike wildfire prevention, fuel treatments are largely conducted by the private sector, on private lands, making it difficult to coordinate an optimal response. Therefore, we explore how changes in (1) the timing of prevention activities, (2) the prevention budget, and (3) both the timing and budget of prevention can be used to reduce the cost-plus-loss of wildfire management, all while assuming a fixed level of prescribed fire treatment.

4 Optimal timing of wildfire prevention effort

We examine how the regions in Florida could time wildfire prevention effort throughout the year to minimize the cost-plus-loss of wildfire management. The timing of the prevention activities are chosen to maximize their effect throughout the wildfire season, while explicitly accounting for the short-run effect of prevention messages (i.e., message effectiveness seems to last six months from the month they were delivered) and the longer-run negative feedback (fuel accumulation) effect. We assume shifts in prevention effort within the year are feasible. We recognize that this may be challenging with fixed staffing levels.

We evaluate two scenarios: (1) assuming a constant prevention budget and (2) allowing for a change in prevention spending. The first scenario explores "what can be done" given the same budget. The second scenario explores "what could be done" with budget flexibility. Both scenarios are run assuming either that (a) individual regions can change their spending patterns across months differently, or that (b) all regions must change their spending patterns equally, by the same percentage across all months, statewide. In addition, we estimate the economic impact of each scenario by estimating the associated cost-plus-loss incurred from changes in wildfire prevention effort.

4.1 Budget: no change

Figure 4 presents the change in wildfire prevention effort for the state (overall) and for the four regions, for the fire years of 2003 to 2007, given no change in the current budget. The model was constrained so that the largest reduction in spending allowed in any month was 99% compared to base level spending. As can be seen, large increases in prevention activity are warranted from January through March, at the expense of activity in the summer and at the beginning of the winter. All regions require large increases in February and March, with Regions 2, 3, 4 also requiring large increases in January. Unlike the other regions, it is more effective to target ignitions later in the year (April) than earlier (January) in Region 1. This result may be linked to the ecological and climatic differences of that part of the state relative to others.

Economically, altering the monthly timing of wildfire prevention activities is expected to yield statewide net benefits of $3.9 million, for the scenario when all regions are required to change their spending by the same amounts (see table 3). When allowed to change independently at different rates, the net benefits increase to $4.3 million over the five years in total. The largest net benefits

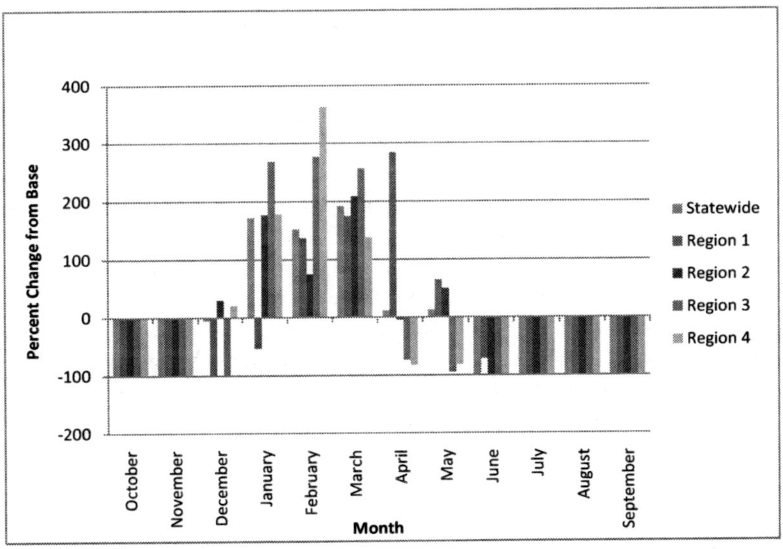

Figure 4: Optimal timing of wildfire prevention effort with no change in prevention spending, statewide and individually by region.

Table 3: Economics of altering the timing of wildfire prevention effort with no change in prevention spending (shown in thousands of U.S. dollars), totals summed over 2003 to 2007.

	Statewide	Region 1	Region 2	Region 3	Region 4
Base spending	2,228	557	557	557	557
Total spending change	0	0	0	0	0
Rx fire spending	12,850	5,796	1,755	2,769	2,531
Wildfire losses	306,857	17,209	68,455	28,178	192,623
Total cost + loss	321,935	23,561	70,767	31,503	195,711
Base cost + loss	325,825	23,673	74,327	31,776	196,049
Change in cost + loss	-3,890	-112	-3,560	-272	-338

would accrue to Region 2, at $3.6 million. Again, these returns occurred from just altering the timing of prevention messages and could be done without any additional spending. While such spending adjustments may involve hidden costs that we cannot account for, at least for the costs that we can account for, the benefit-to-cost ratio implied by such a change is essentially infinite. This demonstrates how important it is for prevention messages to get out ahead of the forthcoming wildfire season.

4.2 Budget: increase

Figure 5 presents the change in wildfire prevention effort for the state (overall) and for the four regions, for the fire years of 2003 to 2007, given an increase in prevention spending. The size of the spending increase was determined by minimizing the cost-plus-loss of wildfire management while allowing for increases in wildfire prevention spending. As in the fixed budget scenario, the simulation was constrained so that the largest reduction in spending allowed in any month was 99% compared to base levels. Note, the overall timing pattern is similar to the one with a fixed budget, with January through March receiving most of the increases. We find that the size of the overall increases are larger than with the fixed budget. The largest monthly increases range from 152% to 192% in the fixed budget case, but from 253% to 331% in the increased spending case. Even though increases in the prevention budget are justified on economic grounds, these increases are targeted to certain times of the year. We find little support for large prevention campaigns in June through October, regardless of budget.

Economically, altering the monthly timing of wildfire prevention activities with greater funds available to support an expansion wildfire prevention effort yields statewide net benefits of $4.4 million (see table 4). Again, the largest net benefits would accrue to Region 2, at $3.7 million. Supporting such economic returns would require an additional $1.0 million of prevention funding over the 2003 to 2007 period if equal changes were imposed across all regions. This represents a 45% increase in wildfire prevention spending under the statewide scenario and 34% when individual regions are allowed to change their spending

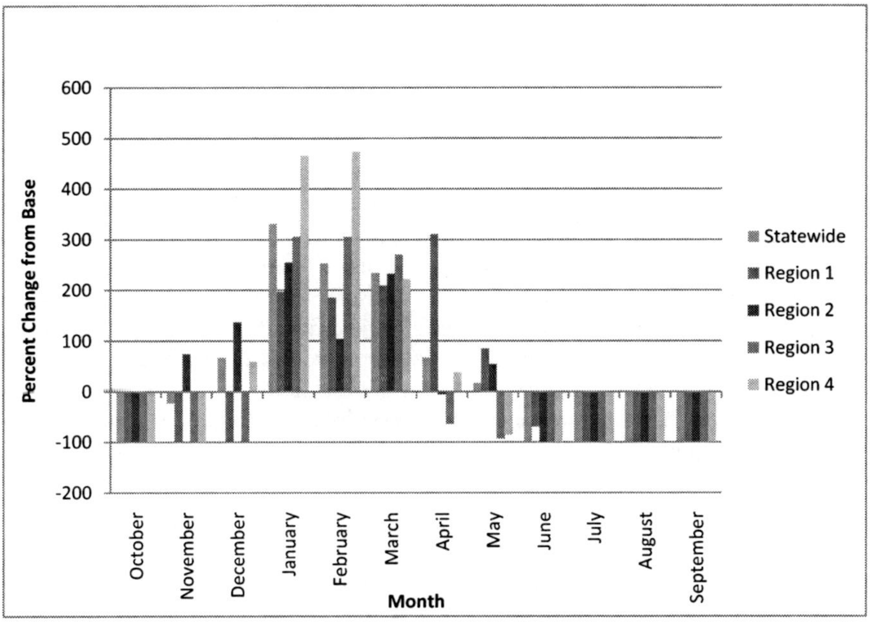

Figure 5: Optimal timing of wildfire prevention effort with an increase in prevention spending.

Table 4: Economics of altering the timing of wildfire prevention effort with an increase in prevention spending (shown in thousands of U.S. dollars), totals summed over 2003 to 2007.

	Statewide	Region 1	Region 2	Region 3	Region 4
Base spending	2,228	557	557	557	557
Total spending change	1,001	187	204	44	312
Rx fire spending	12,850	5,796	1,755	2,769	2,531
Wildfire losses	305,344	16,974	68,104	28,131	192,092
Total cost + loss	321,424	23,513	70,620	31,501	195,492
Base cost + loss	325,825	23,673	74,327	31,776	196,049
Change in cost + loss	-4,401	-160	-3,707	-275	-557

patterns differently. While this increase may be seen as significant, the $1.0 million directly returns $5.4 million in reductions in wildfire losses. This return represents a 5.4 benefit-to-cost ratio. When individual regions are allowed to change differently, the net benefits are even higher, with additional costs over the 5-year period totaling $0.75 million, yielding a reduction in losses of $5.4 million, implying a benefit-to-cost ratio of 7.3.

5 Conclusion

We have shown that the benefits of wildfire prevention and education activities far outweigh their costs. However, the return on investment varies—i.e., effectiveness of these programs—with many exogenous factors, including weather, recent wildfire history and fuel management activities, and season. Given this, we explored the optimal timing of wildfire prevention activities that yield the least wildfire-induced cost-plus-loss to society. We find that, for the State of Florida, changes in the monthly timing of wildfire prevention activity can pay dividends to society. Simply shifting prevention activities to occur during more winter months can produce net benefits of \$3.6 million (assuming similar annual trends [e.g., weather] as occurred from 2003 to 2007). Increases in wildfire prevention spending, of 34% to 45%, coupled with the expansion of prevention effort in the winter and spring months are estimated to produce \$4.4 million in net benefits.

References

[1] Butry, D.T., Fighting fire with fire: estimating the efficacy of wildfire mitigation programs using propensity scores. *Environmental and Ecological Statistics*, **16**, pp. 291-319, 2009.

[2] Pollet, J., & Omi, P.N., Effect of thinning and prescribed burning on crown fire severity in ponderosa pine forests. *International Journal of Wildland Fire*, **11(1)**, pp. 1-10, 2002.

[3] Mercer, D.E., Prestemon, J.P., Butry, D.T., & Pye, J.M., Evaluating alternative prescribed burning policies to reduce net economic damages from wildfire. *American Journal of Agricultural Economics*, **89(1)**, pp. 63-77, 2007.

[4] Mercer, D.E., & Prestemon, J.P., Comparing production function models for wildfire risk analysis in the Wildland-Urban Interface. *Forest Policy and Economics*, **7(5)**, pp. 782-795, 2005.

[5] Prestemon, J.P., & Butry, D.T., Time to burn: modelling wildland arson as an autoregressive crime function. *American Journal of Agricultural Economics*, **87(3)**, pp. 756-770, 2005.

[6] Prestemon, J.P., Butry, D.T., Abt, K.L., & Sutphen, R., Wildfire Prevention Efficacy: Marginal and Non-marginal Benefit to Cost Ratios. *Forest Science*, In press.

[7] Butry, D.T., Prestemon, J.P., Abt, K.L., & Sutphen, R., Economic optimization of wildfire intervention activities. *International Journal of Wildland Fire*, In press.

[8] Butry, D.T., Pye, J.M., & Prestemon, J.P., Prescribed fire in the interface: separating the people from the trees. *Proc. of the Eleventh Biennial Southern Silvicultural Research Conference*, ed. K. Outcalt, Gen. Tech. Rep. SRS-48, USDA Forest Service, Ashville, pp. 132-136, 2002.

[9] Bradshaw, W.G., Fire protection in the urban/wildland interface: who plays what role? *Fire Technology*, **24(3)**, pp. 195-203, 1988.

Comparing environmental values across major U.S. national parks

D. B. Rideout, P. S. Ziesler & Y. Wei
Fire Economics and Management Laboratory,
Department of Forest, Rangeland and Watershed Stewardship,
Colorado State University, USA

Abstract

Environmental values at four of America's most famous national parks are assessed through a structured survey technique. Resource and wildland fire managers in Sequoia and Kings Canyon, Yellowstone, Grand Canyon, and Rocky Mountain national parks were surveyed to identify the most important values that could be improved or that should be protected from wildland fire. In some instances wildland fire can produce beneficial effects on the landscape by treating vegetation to restore or maintain the natural system. In other circumstances, wildland fire can severely affect life, property and treasured natural resources. This research identifies the key sets of values potentially affected by wildland fire across each of the four national parks and compares the relative importance of each kind of value. With a diverse set of famous North American national parks, this research shows which values they share and which values are unique. This was accomplished by implementing a structured and peer-reviewed elicitation process. For the first time, we show the set of values important to each of these national parks and how they compare and contrast.

Keywords: Grand Canyon, Yellowstone, Rocky Mountain, Sequoia Kings Canyon, valuation, nature's services, environmental capital, meta-value, disturbance, wildland fire.

1 Introduction

America's national parks contain some of the most treasured natural resources in the country and the world. While each national park is unique in its natural features and unique in its enabling purposes and legislation, all parks share a

WIT Transactions on Ecology and the Environment, Vol 137, © 2010 WIT Press
www.witpress.com, ISSN 1743-3541 (on-line)
doi:10.2495/FIVA100191

common sense of environmental purpose and value. For example, some parks contain unique forest cover types, such as the famous sequoia trees in southern California, while others are home to critical wildlife habitats, such as the northern Goshawk. However, given the mission of national park management, we would also expect the parks to exhibit some similarities in valuation. Exploration of valuation across America's national parks has not previously been approached using a quantitative framework. Fundamental questions of interest for understanding the importance of nature's (natural) services and the importance of environmental capital include:

- Can unique national parks be characterized by a shared set of meta-value attributes (shared environmental values)?
- If they can be characterized by common meta-value attributes, how do the individual attribute values compare across the diverse parks?

The first question seeks to identify if there are common categories of value shared among the parks. For example, while there are unique cover types and unique wildlife species, can workable general categories be constructed that can then be used to address valuation across parks? The second question is contingent upon a positive response to the first. For example, if meta-value categories can be formed that apply across the parks, we would want to know if and where the values show consistency and where they show differences. These questions are of general interest in environmental and resource management and they can provide valuable insight and information for resource planning and environmental compliance. This paper will address the characterization and comparison of values for national parks in the context of disturbance management – particularly the management of wildland fire. Value information for disturbance management is broadly useful because it can be employed for planning exercises and compliance and it may be used in a general context for evaluating resource value changes from disturbances.

Like all natural systems, national parks are subject to natural disturbances such as hurricanes, insect infestations and wildfire. A consequence of disturbances is that they often inflict a wide array of resource changes [1] and a corresponding array of changes in resource values. In wildland fire management, disturbances usually affect multiple resources where some resources are improved and others may be damaged. For example, a single low-intensity wildland fire might have a positive impact on cultural resources such as Giant Sequoia groves and a negative impact on habitat for sensitive wildlife. Capturing the directions and relative magnitudes of such value changes is critical for evaluating potential management responses to a fire disturbance.

Valuation for disturbance management typically uses marginal values rather than total values because both management activities and disturbances usually produce marginal changes (modest increases or decreases) to the level of natural services and/or environmental capital stock. Our valuation comparisons rely on the ideas that 1) resource value changes are marginal or incremental and 2) that the marginal values are measured on the same numeric cardinal scale (they are relative values). The data set used to estimate fire value comparisons among the four national parks includes a list of resource values affected by wildland fire and the relative change in value (marginal value) induced by wildland fire under specified conditions.

In 2008, the U.S. federal government spent approximately $3.0 billion on wildland fire suppression and related activities and another $3.25 billion [2] on fuel treatment. In some federal agencies, including the U.S. Forest Service, expenditures on managing fire disturbance now comprise a majority of the agency budget. Information on resource value changes related to disturbance events, including wildland fire has, perhaps, never been so timely. Because forest disturbances affect the flow of nature's services and the value of the underlying environmental asset, a clear and pragmatic understanding of the value changes involved can promote viable levels of natural services and corresponding asset values. Understanding such value changes can assist resource managers by aiding in selection of the most appropriate intervention(s) and by guiding the placement of mitigation actions, such as fuel treatments, on the landscape. Such information is highly useful for assessing land management planning and in addressing specific loss mitigation strategies.

The paper is organized as follows: after a brief review of previous work in resource and disturbance valuation we discuss the elicitation of resource value attributes and their values for each national park. We used the MARS elicitation method [3] on each park to establish the pertinent list of resource values affected by fire and the marginal values for each resource under various ecosystem conditions and fire intensities. The results and discussion and conclusions sections provide the comparisons and contrasts of the resulting valuation data.

2 Previous work

Since the publication of Paul Samuelson's work on pubic goods [4, 5], resource economists have laboured to develop the theory and techniques of resource valuation. Today's formal techniques for assessing environmental values are intensive, costly and primarily resource specific. Most of the accepted techniques reflect enormous advances since Samuelson's work, but they are not readily applied to the wide array of resources affected by wildland fire. Consequently, resource managers and planners are saddled with the problem of generating resource disturbance values for strategic, long-term program planning, and for rapid response planning when a disturbance strikes. Fire managers are routinely required to make resource management decisions involving the marginal valuation of a wide range of natural resources. They are required to make mitigation or enhancement decisions quickly when there is an increment of value to add or to protect by managing a fire event. Most fires, including prescribed burns, will promote modest changes to the fuel conditions on a landscape. Such an incremental change is often appropriately addressed through "marginal analysis".

Since the work of Samuelson, resource economists have focused on applying principles of demand theory to address the valuation problem and they have been careful to distinguish between total value (the value of a defined quantity enjoyed), marginal value (the change in total value resulting from small changes in quantity) and consumer surplus (the total consumer benefit minus payment). While most studies and applications of valuation theory address total value and consumer surplus and thereby enable an estimate of marginal value [6], few directly estimate marginal values; especially across the set of resources impacted by disturbance.

A recent survey of modern valuation techniques [7] explained the use of an array of "revealed preference" and "stated preference" techniques. In addition, the field of "experimental economics" is expanding upon the application of stated preference theory through structured experimental design [8]. Revealed preference techniques rely on observing choices that resource users (or consumers) make. Revealed preference techniques include the classic travel cost method, where expenditures on travel are observed and analyzed to estimate outdoor recreation values. While revealed preference techniques have the advantage of using observed behaviour, they are impossible to use where valuation of natural systems involves non-use values. Non-use values include concepts such as a society's value for the ability to transfer natural treasures to future generations or an individual's value of knowing that a resource exists. Stated preference methods do not rely on observed choices but have the important advantage of being able to address non-use public good values such as the value of preserving resources for future generations, while not sacrificing the ability to address use values. Stated preference techniques have greatly advanced in recent years and include a variety of contingent valuation methods and a family of hedonic approaches [7]. Despite such great strides in theory and techniques, land managers continue to face an important scarcity of information on marginal values. Much of the valuation information that is available pertains only to a specific resource condition. This specificity prevents it from being reliably applicable to other resources or conditions. Further, some estimates are too general to be credible at the planning unit level. Techniques are needed that directly apply to the wide range of resources potentially affected by disturbances and that are appropriate for the local planning unit level.

To address the issue of pragmatic relevance of resource values for fire management planning, Rideout *et al.* [3] developed an approach to estimate marginal relative values at the planning unit level. The approach, known as "Marginal Attribute Rates of Substitution" (MARS), requires a structured elicitation of values from fire and resource management officials at the local planning unit level. MARS was successfully applied at the Sequoia and Kings Canyon National Parks (SEKI) in May 2005. The MARS process required about two days of intense elicitation and the information supplied through MARS was used to inform fire management and firefighting efforts on the SEKI landscape during the 2008 fire management season.

After SEKI, MARS was applied at Grand Canyon (GRCA), Yellowstone (YELL) and Rocky Mountain (ROMO) parks yielding a four-park data set of marginal valuation information that has not previously been analyzed. The four parks include famous national treasures and each has important fire and resource management issues requiring information on marginal valuation. The addition of these three parks provides an interesting set of valuation data that addresses important differences among the ecosystems and the extent that values across the ecosystems are similar or different.

3 Data sources and methods

Rideout *et al.* [3] and others found that the valuation problem for disturbances such as wildland fire differs different from direct valuation of the resource. Because value changes associated with disturbances are rarely collected or documented in land and resource management planning processes, marginal valuation estimates for fire management were elicited from managers at each of the four national parks. While each national park is unique, there are important similarities among them with respect to fire management and valuation as shown below.

3.1 The parks

Sequoia and Kings Canyon National Park (SEKI), located in the Sierra range in southern California, contains the unique and famous sequoia groves including “named trees” of special importance such as the “General Sherman” tree and the Grant Grove. Sequoias are a fire-adapted species with thick bark to protect them from low-intensity fire events. The fire return interval for sequoias is relatively short at approximately 17 years. Frequent low-intensity fires in the groves provide benefits by clearing out encroaching vegetation and removing fuels that could ultimately produce a catastrophic event. The park is also home to other forest cover such as the “mixed conifer” type. Mixed conifer is known for its longer fire return interval, thin bark and abundant ladder fuels that enable fires to travel up trees into the crown where they can increase in intensity and rapidly spread. SEKI is also home to ponderosa pine. Although not unique like sequoia, ponderosa pine with its thick bark shares many fire behaviour and effect characteristics with the sequoia. Grand Canyon National Park located along the Colorado River in Arizona and Utah shares some similarities with SEKI. Both parks have high-value cultural resources that are affected by fire and management activities. There are also important differences between the parks in that GRCA does not have the same topographic profile as SEKI and it is home to tree types and wildlife not present in SEKI, such as the pinion-juniper (PJ) cover type and the goshawk.

In contrast, Rocky Mountain National Park (ROMO) and Yellowstone National Park (YELL) are located in the high elevation Rocky Mountains, are known for their long fire return interval cover types and contain different fire-adapted species than SEKI or GRCA. ROMO is located in northern Colorado and contains the headwaters of the Colorado River that ultimately flow through GRCA. YELL is located in northwestern Wyoming and is famous for its geothermic sites and variety of wildlife. Similar to SEKI and GRCA, these parks contain mixed conifer type, but they also have vast stands of lodgepole pine. The lodgepole pine, while fire adapted, responds differently than the sequoia or the ponderosa. Fires in lodgepole pine are often referred to as "stand replacing" because entire stands are designed to be well consumed and then regenerated by wildfire. Individual trees reproduce by shooting seed from their cones when activated by heat. Such similarities and differences in vegetation types between the parks make for informative pairwise comparisons.

An interesting condition of the short fire return interval ecosystems is that while the desired condition can quickly degrade, it can also readily be returned to a more desirable state by using fire as a treatment. When these systems do not experience fire consistent with their natural return interval (they miss a return interval by chance or by an active fire suppression program), they begin to change in undesirable ways. With missed intervals they accumulate fuels and other understory and tree species may "invade". Here, stand and ecosystem conditions can benefit from reintroducing fire through fuel treatments or by allowing ignitions to play their natural role. In contrast, cover types with longer fire return intervals are typically considered to be in a "steady state" of ecosystem maintenance. While they can benefit from fire, the amount of time required for departure from the natural condition is lengthy. Hence, they will not register positive marginal values for fire in the same way as the short return interval parks.

3.2 Value elicitation using the MARS method

Application of MARS [3] to each park entailed assembling a full set of fire management planners to form an "expert" group. This group included resource management specialists, cultural resource specialists, fire management operations specialists and land management planners. Given the direction from their land management and fire management plans they identified the list of natural resources (known as value attributes) they actively manage for positive and negative fire effects. First, each expert group identified a set of value attributes appropriate to their park. Value attributes are a specific type of resource affected by fire such as high value development areas, sequoia groves (SEKI only) or lodgepole pine cover type (YELL, ROMO). Cover types are often a proxy for a set of resources associated with the cover. Next, as appropriate, value attributes are further categorized by ecosystem condition and fire intensity. Condition and intensity categories for some value attributes are necessary because fire effects or their importance often differ by ecosystem condition and/or fire intensity. For example, sequoia groves and ponderosa pine stands typically benefit from low-intensity fire when ecosystems are in a "maintenance mode" (no missed fire intervals) while stands and ecosystems might be harmed by high-intensity fire regardless of management mode. MARS was specifically designed to admit the negative and the beneficial effects of fire; in particular, it permits managers to capture the management concept that low-intensity fires may be used in certain cover types to maintain the system in a desirable state. Once the collection of value attributes is elicited and categorized by condition and intensity, the entire list is carefully reviewed by the group to ensure it is complete, appropriate for strategic planning, and that it does not include any double counting (identifying the same resource value under two headings). The final part of the elicitation process estimates the marginal values of each attribute under each intensity and ecosystem condition. These marginal values are known as marginal rates of substitution (ROS) in the economics literature. They indicate the rate at which one value can, in principle, be substituted for any of the others. A value of -1.0 defines the most important resource to protect (per hectare) from fire and a value of -0.5 defines a resource and condition that is half as important. Positive values, from 1.0 to 0.0 denote beneficial fire effects. They are interpreted similarly and are symmetrical with the negative values. These rates of substitution are the key valuation information produced from the MARS process and because they use a common currency, they are comparable across planning units.

4 Results and discussion

The results for value attribute categories and their rates of substitution are addressed starting with "human" resources such as developments and then natural designations such as wildlife. Values were elicited for the management mode "maintenance" at high and low fire intensities at all parks, and for the management mode "restoration" at high and low fire intensities for SEKI and GRCA.

The four parks have high value development that serves as a common currency at -1.0 for each combination of fire intensity (high and low) and management mode. These are of the highest importance to protect from fire. The value attribute "sensitive boundary" appears in GRCA, YELL and ROMO and under all modes and intensities it was very high for each of these parks. YELL had an elicited value of -1.0 for all four combinations of mode and intensity and GRCA had -0.80 for all four combinations. ROMO only has values for restoration mode at its boundary and the elicited values are -1.0 for high and low fire intensities. Protection of development and park boundaries, particularly boundaries near the Wildland-Urban Interface, are expected to have high importance to protect from fire; so the presence of the development and boundary attributes in three of the four parks and the elicited values associated with them at or near -1.0 are as expected.

Archaeological and cultural sites appeared in the attribute lists of three national parks: SEKI, GRCA and ROMO. Yellowstone did not have a separate attribute for cultural sites apart from "high value development". For the three parks with this attribute, all elicited values were negative, indicating that fire has the potential to damage such sites and therefore they are important to protect from fire. Under all combinations of management mode and fire intensity, the values ranged from -0.65 to -0.95. SEKI had the lowest values (-0.65) for low intensity fires under both management modes and -0.80 for high intensity fires (both management modes). GRCA had the same value, -0.85, for all four combinations of intensity and management and ROMO had the highest value, -0.95, for both high and low intensity fires under maintenance (no values for restoration mode). The differences among the parks are not unexpected for this attribute because of differences in expected fire behaviour at the archaeological and cultural sites for each park.

Each of the four parks has different wildlife, some of which have fire-sensitive habitats. Two parks identified wildlife habitat areas that are affected by fire management: ROMO identified elk winter habitat and GRCA identified goshawk and Mexican spotted owl habitats. Each park elicited values for individual species, but there are similarities among the values even though the species are very different. All values for low intensity fires, whether under maintenance or restoration, were positive. This indicates that both parks consider low intensity fire as beneficial for selected wildlife. Importance to improve values for low intensity fires ranged from 0.40 for Mexican spotted owl habitat to 0.55 for goshawk habitat. ROMO was in the middle-high range with 0.50 for elk habitat. For high intensity fires, such values were negative. The importance to protect values at high intensity have a wider range than the importance to improve values with ROMO showing an importance to protect value for elk habitat of -0.1 and GRCA showing -0.4 for Mexican spotted owl habitat and -0.5 for goshawk habitat. All except elk winter range at high intensity are mid-way between the extreme values of 0 and +/-1.0. An interpretation of the importance to protect habitat values would be that they are approximately half as important to protect as high value developments during high intensity fires.

Elicited values for short-return fire interval forests are similar in spite of being complex. SEKI and ROMO have ponderosa pine attributes and SEKI has a sequoia attribute. The only combination of intensity and mode that provides a comparison between parks is the low-intensity maintenance mode. For both parks, short-return fire interval forest types in maintenance mode have a high benefit from low-intensity fire events: sequoia groves have a value of 1.0 and ponderosa pine (both parks) has a value of 0.90. It is difficult to compare the parks under the *other* combinations of intensity and mode because ROMO shows only one other value, the high-intensity maintenance mode combination (value of 0.50), and SEKI does not have this combination. Instead, SEKI has values under restoration reflecting negative effects at high-intensity (-0.6 for sequoia and -0.8 for pine) and benefits at low-intensity (0.8 for sequoia and 0.7 for pine). When viewed as a group, all short return interval values are positive except those under restoration for high-intensity fires. High intensity fires in stands under restoration would be harmful while other fires are beneficial.

Three national parks list attributes and values for long-return interval forest cover types. YELL and ROMO have lodgepole pine and spruce-fir, and SEKI and ROMO have a mixed-conifer type. Only the SEKI mixed-conifer is in restoration mode, with values of -0.3 for high-intensity and 0.6 for low-intensity. All other long-return interval cover types for these parks were considered to be in maintenance mode. Among the parks with long-return interval cover types, very similar values were elicited for importance to improve with fire when the stands are in maintenance mode. All of the elicited values for both high and low fire intensity are 0.8, except for lodgepole pine in ROMO at low fire intensity (0.70).

Table 1: Fire management meta-value attributes for four national parks.

Short Fire Return Interval Cover Types	**Cultural and Archaeological Sites**
Sequoia (SEKI)	Cultural Trees (SEKI)
Ponderosa Pine (SEKI, ROMO)	Cultural Sites (ROMO)
	Sensitive Archaeological (GRCA)
Long Fire Return Interval Cover Types	
Spruce-Fir (YELL, ROMO)	**High Value Development**
Lodgepole Pine (YELL, ROMO)	(All Parks)
Mixed Conifer (SEKI, ROMO)	
	Wildlife Habitat
Sensitive Boundary	Goshawk Nesting (GRCA)
(GRCA, YELL, ROMO)	Mexican Spotted Owl (GRCA)
	Elk Winter Range (ROMO)

Examination of the attribute lists and elicited values from the four national parks suggests that construction of a set of meta-value attributes would be promising. Definition of a meta-value attribute appears straightforward when multiple parks have identified the same value attribute, such as "high value development". Unique values, such as the value of protecting volcanic monitoring sites in YELL, are not useful for the meta-value list because they do meaningfully compare or contrast with attributes in other parks. Other unique attributes may be compared across parks as they fit broader categories. An example is combining goshawk nesting sites and Mexican spotted owl habitat in GRCA with winter elk habitat in ROMO to define a "wildlife habitat" meta-value attribute. If such meta-value attributes are viable, they have the potential to suggest resource values for other planning units and reference points in marginal valuation for other parks. Analysis of the four park data sets provides a list of park meta-value attributes. Table 1 lists the meta-value attributes (bold) and their specific value attributes as defined by the parks.

The unique features of U.S. national parks suggest a range of values for each meta-value attribute. Figure 1 shows the ranges and midpoints graphically to enhance the comparisons of relative values and their ranges. Immediately apparent is the range of values for wildlife habitat under restoration and high intensity fire. This is the highest range in the figure at 0.4 values units, or 20% of the overall range. This is not surprising as high intensity fires often have a wider range of impacts. Five of the 16 categories have the narrowest range of 0.0 units. These are high value development, ponderosa/sequoia for either management mode at low fire intensity, long return interval conifers in maintenance mode at low fire intensity and wildlife habitat in maintenance mode at high fire intensity.

Resources designated, created and maintained by humans such as developments, boundaries and cultural and archaeological sites, tend to have high importance to protect from fire. The range of these values is between -0.65 and -1.0. Resources created and maintained by nature tend to have importance to improve, particularly at low fire intensities or when in maintenance mode for management. Only when nature's resources are in a restoration mode or when high fire intensity may damage a resource do they incur an importance to protect from fire. Unlike the human created resources, nature's meta-value attributes did not follow a consistent result of "always protect" or "always improve".

Although for a *given* attribute and management mode/fire intensity combination, either "always protect" or "always improve" may be applied. Importantly, none of the meta-value attributes under any mode/intensity combination have a range that crosses zero. This suggests that meta-value attributes may be useful descriptors of a shared set of environmental values across at least the four national parks evaluated.

5 Conclusions

The results provide a means of addressing the two fundamental questions posed for this paper: can potentially useful meta-value attributes be constructed for

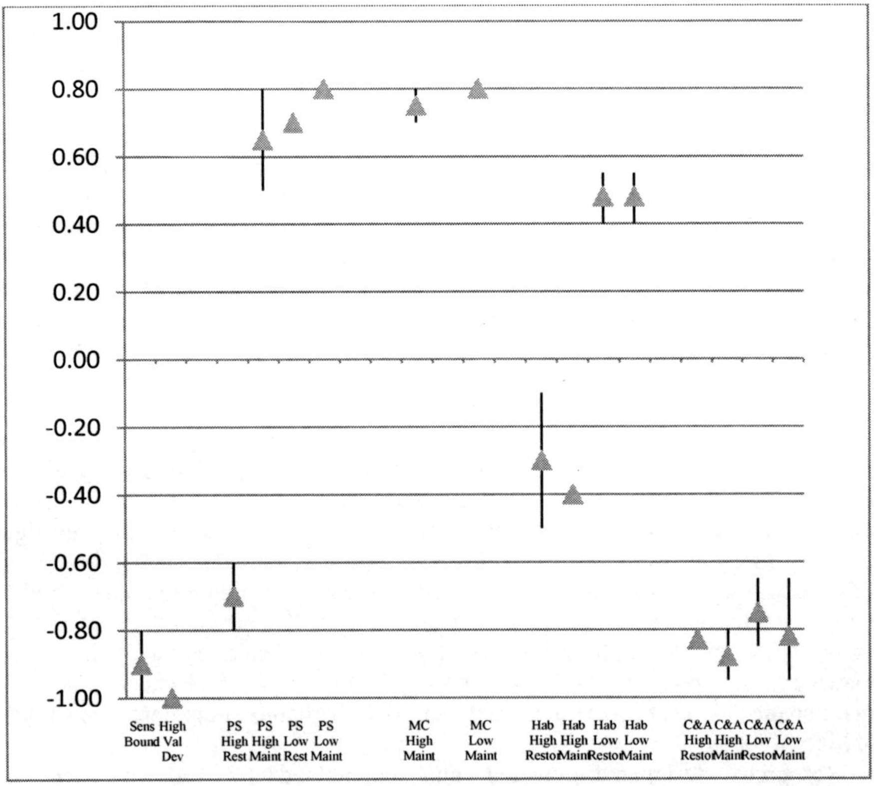

Figure 1: Ranges and mid-points for meta-value attributes. PS denotes ponderosa and sequoia, MC denotes mixed conifer, Hab denotes wildlife habitat and C&A denotes cultural and archaeological sites.

U.S. national parks and how would those values compare. Interpretation of the results suggests that meta-value attributes are a viable construct to enable broader interpretation of marginal rates of substitution. For cover types, this is greatly facilitated by distinguishing between long and short fire return interval systems. While we express caution considering that only four parks were sampled, our descriptive data suggest that for most meta-value attributes there is evidence of consistency in valuation across these western U.S. national parks. Given that MARS was independently applied using four different "expert" groups, the consistency suggests that the results of MARS may be largely independent of the particular expert group. The consistency of values is shown by the tight ranges produced across the parks and by the fact that none of the ranges extended across zero. To the extent that the meta-values are viable and consistent, they can be used on other parks to facilitate the valuation process. Parks initiating the marginal valuation process can use the meta-value attributes to better identify their own specific planning unit level attributes. Public officials charged with management of treasured natural resources in the western U.S. parks exhibited a high level of consistency across fire intensities and ecosystem conditions.

References

[1] Holmes, T. P., Prestemon, J. P. & Abt, K. L., An Introduction to the Economics of Forest Disturbance (Chapter 1). *The Economics of Forest Disturbances: Wildfires, Storms, and Invasive Species*, ed. T. P. Holmes, J. P. Prestemon & K. L. Abt, Springer Science + Business Media B. V., pp. 3-14, 2008.

[2] U. S. Senate, Department of the Interior, Environment, and Related Agencies Appropriations Bill, 2008.

[3] Rideout, D. B., Ziesler, P. S., Kling, R., Loomis, J. B. & Botti, S. J., Estimating rates of substitution for protecting values at risk for initial attack planning and budgeting. *Forest Policy and Economics*, **10**, pp. 205-219, 2008.

[4] Samuelson, P. A., The Pure Theory of Public Expenditure. *Review of Economics and Statistics*, **36**, pp. 387-389, 1954.

[5] Samuelson, P. A., Diagrammatic Exposition of a Theory of Public Expenditure. *Review of Economics and Statistics*, **37**, pp. 350-356, 1955.

[6] Loomis, J. B., *Integrated public lands management: principles and applications to national forests, parks, wildlife refuges, and BLM lands, Second Edition*, Columbia University Press: New York, NY, 2002.

[7] Champ, P. A., Boyle, K. J. & Brown, T. C., (eds). *A Primer on Nonmarket Valuation*, Kluwer Academic Publishers: Dordrecht, The Netherlands, 2003.

[8] Davis, D. D. & Holt, C. A., *Experimental Economics*, Princeton University Press: Princeton, N.J.: 1993.

Section 6
Risk and vulnerability assessment

A volatile organic compounds flammability approach for accelerating forest fires

L. Courty[1,2], K. Chetehouna[2], J. P. Garo[1] & D. X. Viegas[3]
[1]*Institut P', UPR 3346 CNRS, ENSMA, Futuroscope Chasseneuil, France*
[2]*ENSI de Bourges, Institut PRISME, Bourges, France*
[3]*ADAI, Department of Mechanical Engineering, University of Coimbra, Portugal*

Abstract

In this paper the accelerating forest fire phenomenon for three real accidents is investigated. This phenomenon is studied using the hypothesis of the ignition of a Volatile Organic Compounds (VOCs) cloud accumulated in canyons. By heating a *Rosmarinus officinalis* plant in a specific hermetic enclosure, fourteen VOCs are identified and quantified as temperature functions. The theoretical flammability limits of those components are calculated by means of empirical correlations. Froude scaling law is applied to present laboratory results to find the concentrations of VOCs at canyon scale. The comparison of the flammability limits with the obtained concentrations shows that the emitted VOCs can lead to an accelerating forest fire.

Keywords: Rosmarinus officinalis, VOCs emission, flammability domain, Froude scaling law, accelerating forest fire.

1 Introduction

Forest fires cause important damages in terms of ecological and economical issues and human lives every year. Some fires with normal behavior suddenly start to propagate at unusual and very fast speed. They are called eruptive fires by Viegas [1], fires flashover according to Dold et al. [2] or accelerating forest fires by Chetehouna et al. [3]. The term eruptive refers to the continuous rise of the fire rate of spread and flashover to its unexpected character.

There have been many accidents over the last half century where eruptive fires were reported [4–8]. In France, sixteen firefighters were killed during the

WIT Transactions on Ecology and the Environment, Vol 137, © 2010 WIT Press
www.witpress.com, ISSN 1743-3541 (on-line)
doi:10.2495/FIVA100201

last fifteen years because of this phenomenon [9]. The more investigated case is the accident of Palasca (Corsica Island) in 2000, where survivors declared to have been surrounded by a "lake of fire" [2]. In Portugal, more than fifty persons lost their lives since 2003 [10, 11]. More recently, the Kornati accident in Croatia in 2007 killed twelve persons [7]. In Greece, also in 2007, more than seventy five persons lost their lives [12].

Up to now, the mechanisms to explain this kind of forest fire are not totally well known, as there are several phenomenological approaches for this phenomenon in the forest fires literature. One of the most recent mechanisms, proposed by Viegas [1, 13], suggests that the "fire eruption" is the consequence of the feedback effect of the convection induced by the fire. Another approach is that the acceleration of the rate of spread is caused by flow attachment in the direction of the fire propagation, according to Dold and Zinoviev [14].

In this paper, we explore the possibility of explaining the sudden acceleration of the fire as a consequence of the inflammation of the Volatile Organic Compounds (VOCs) emitted by fire heated vegetation and accumulated in the terrain ahead of the fire front for specific geometrical configurations such as canyons. Indeed, Chetehouna et al. [3] have shown using a simple extrapolation from laboratory experiments that the concentration of the VOCs at field scale is in the flammability domain (i.e. between the Lower Flammability Limit and the Upper Flammability Limit) of the α-pinene, which is the main constituent emitted by *Rosmarinus officinalis* plants.

The aim of this paper is to investigate more deeply this hypothesis by comparing the concentration of the VOCs mixture at canyon scale obtained by Froude scaling law for three accidents with the computed flammability limits. This approach assumes that the canyon is located in a place with a wind lull so that the VOCs accumulation is not modified by the atmospheric wind nor by the one induced by the flames. The sudden inflammation of these VOCs can possibly happen due to a firebrand coming from the fire front.

The second section is devoted to the description of the experimental protocol and to the presentation of the VOCs emissions at laboratory scale. In the third one, we will use an empirical method to determine the flammability limits of pure compounds at a reference temperature. The dependence in temperature is obtained by applying two correlations to those flammability limits. The flammability domain of the VOCs mixture is given by Le Chatelier's law. The last section is dedicated to the derivation of a relationship giving the VOCs concentration at canyon scale by means of Froude scaling law. Using a geometrical characteristic of the canyon shape, we will compare the VOCs concentrations of three real accidents (South Canyon, Palasca and Kornati) with the flammability domain provided in the previous section.

2 Volatile organic compounds emission in a hermetic enclosure

In this work, about thirty *Rosmarinus officinalis* plants were used to determine the effects of plant temperature on the emission of VOCs. These plants were placed in a hermetic enclosure and were heated by a radiant panel. The hermetic enclosure with the dimensions of $100 \text{ cm} \times 100 \text{ cm} \times 134 \text{ cm}$ was designed in cellular-concrete material with a thickness of 7 cm and its volume was 1.2 m^3. The radiant panel is constituted of 16 black ceramics plates of a 144 cm^2 surface providing a maximal radiative heat flux around $84 \text{ kW} \cdot \text{m}^{-2}$ (Figure 1).

Figure 1: Pictures of the experimental setup.

The average mass, height and moisture content of the *Rosmarinus officinalis* plants were respectively 155 g, 30 cm and 70%. They were placed at the centre of the hermetic enclosure, 50 cm from the radiant panel, and heated during 30 min. The heat flux of the radiant panel was varied from 0.44 to 20.59 $kW.m^{-2}$. The experimental protocol consisted in trapping and sampling the VOCs with glass multibed tubes, to transfer them into a freezing box to the chemistry laboratory and to analyse them with an ATD-GC/MS instrument. Each experiment was performed in triplicate [3].

The aim of these experiments was to study the VOCs emission as a function of temperature in order to estimate the VOCs quantity emitted by the vegetation during a forest fire. The selected temperature range was between 30 and 210 °C to simulate the heating of plants by a fire front before the pyrolysis phase (about 250 °C). The VOCs mixture (14 compounds identified) was characterized by high contents of monoterpenes hydrocarbons. Indeed, the main components were α-pinene, limonene, camphene, myrcene and β-pinene. The same compounds were observed by Ormeño et al. [15] for the study of the VOCs emissions by *Rosmarinus officinalis* under natural conditions. Figure 2 illustrates the evolution of the major constituents and the total VOCs emissions as a function of the plant temperature.

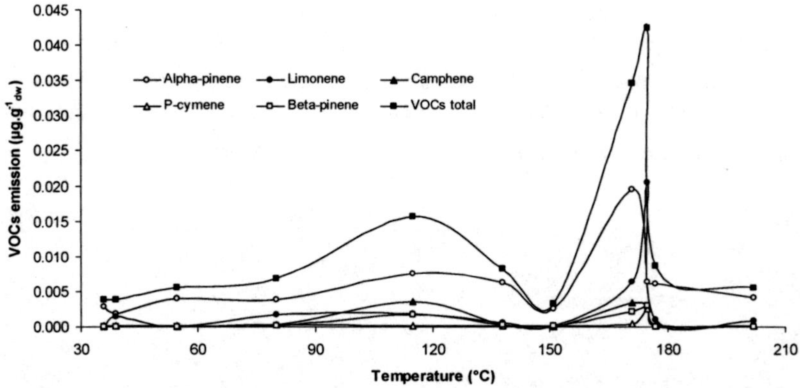

Figure 2: Main and total VOCs emission as function of temperature.

It was noticed from this figure that the VOCs emission increases with the plant temperature until 175°C. The same tendency was observed by Barboni [16] for others Mediterranean tree species such as needles of *Pinus nigra* and *Pinus pinaster*. The VOCs mixture amount at 175°C is 8 times higher than the one measured at 50°C. Moreover, we can see an increase of VOCs production around 120°C due to the transport of the VOCs by the evaporation process. Knowing that the boiling temperature of monoterpenes is about 154°C, these molecules would be in a liquid or in an equilibrium liquid-vapour state below this temperature. As a consequence, for the temperatures higher than this value, the VOCs emission increases rapidly to a maximum at 175°C. In this temperature range, the total emission of the VOCs is about 4 times more important than in the range 50–120°C. After 175°C, we can observe a significant decrease of the VOCs amount that can be explained by the thermal degradation of the terpenic molecules [16].

3 Flammability limits of a volatile organic compounds mixture

As mentioned in the previous section, 14 VOCs emitted by *Rosmarinus officinalis* and their proportion in the mixture have been identified at different temperatures before the pyrolysis phase. In this section, we will calculate the flammability domain of this VOCs mixture.

The lower and upper flammability limits of a VOCs mixture can be expressed by:

$$LFL_{mix}\left[\mathrm{g}\cdot\mathrm{m}^{-3}\right]=10\frac{LFL_{mix}\left[\%\right]}{V_{m}}W_{mix} \tag{1}$$

$$UFL_{mix}\left[\mathrm{g}\cdot\mathrm{m}^{-3}\right]=10\frac{UFL_{mix}[\%]}{V_m}W_{mix} \tag{2}$$

where $LFL_{mix}[\%]$ and $UFL_{mix}[\%]$ are respectively the lower and upper flammability limits in volume percentage, $W_{mix}\left[\mathrm{g}\cdot\mathrm{mol}^{-1}\right]=\frac{1}{\sum_i \frac{y_i}{W_i}}$ is the molar weight of the mixture with y_i being the mass fraction of the ith VOC and $W_i\left[\mathrm{g}\cdot\mathrm{mol}^{-1}\right]$ its molar weight. $V_m\left[\mathrm{L}\cdot\mathrm{mol}^{-1}\right]$ is the molar volume that depends on the temperature as:

$$V_m\left[\mathrm{L}\cdot\mathrm{mol}^{-1}\right]=8.205\times10^{-2}(T+273) \tag{3}$$

where $T[°\mathrm{C}]$ is the temperature of the VOCs mixture.

The flammability limits $LFL_{mix}[\%]$ and $UFL_{mix}[\%]$ can be found with the Le Chatelier's formula [17, 18]:

$$LFL_{mix}[\%]=\frac{1}{\sum_i \frac{x_i}{LFL_i[\%]}},\quad UFL_{mix}[\%]=\frac{1}{\sum_i \frac{x_i}{UFL_i[\%]}} \tag{4}$$

where $LFL_i[\%]$ and $UFL_i[\%]$ are respectively the lower and upper flammability limits in volume percentage of the ith compound of the VOCs mixture and x_i its mole fraction given by

$$x_i=\frac{\frac{y_i}{W_i}}{\sum_i \frac{y_i}{W_i}} \tag{5}$$

Several methods are available in the literature to estimate the lower and upper flammability limits of pure compounds. These methods can be divided into four categories: empirical equations [19–21], critical flame temperature correlations [22, 23], structural group contribution methods [24] and neural network methods [25]. As empirical equations are easy to use (i.e. we need only the molecular formula of the compound) and give relevant results according to the previous authors, we will use them in this paper.

The lower and upper flammability limits in volume percentage $LFL_i[\%]$ and $UFL_i[\%]$ of the ith VOC, at the reference temperature 25°C, are given by the Gharagheizi [20, 21] relations:

$$LFL_{i(25°C)}[\%] = 0.76022 - 3.57754\,PW5 - 1.47971\,AAC$$
$$+8.57528\,SIC0 - 0.01981\,MLOGP \quad (6)$$

$$UFL_{i(25°C)}[\%] = 10.35415 - 1.35486\,Jhetv - 42.28779\,PW5$$
$$+18.59571\,SIC0 + 0.98203\,MATS4m \quad (7)$$
$$-0.68363\,MLOGP$$

The different parameters in these relations are molecular descriptors associated to each VOC and their values are available online on the Milano Chemometrics and QSAR Research Group website. These molecular descriptors are defined in Table 1.

Table 1: Definition of the molecular descriptors.

Molecular descriptor	**Type**	**Definition**
$PW5$	Topological descriptors	Path/walk 5 Randic shape index
AAC	Information indices	Mean information index on atomic composition
$SIC0$	Information indices	Structural information content (neighborhood symmetry of 0-order)
$MLOGP$	Molecular properties	Moriguchi octanol-water partition coefficient (log P)
$Jhetv$	Topological descriptors	Balaban-type index from van der Waals weighted distance matrix
$MATS4m$	2D Autocorrelations	Moran autocorrelation-lag 4 weighted by atomic masses

We can obtain the flammability limits as function of temperature by using equations (6) and (7) coupled with the relations of Arnaldos et al. [26] and Zabetakis [27] respectively:

$$LFL_i[\%] = LFL_{i(25°C)}[\%]\left(1 - 7.8\times10^{-4}(T-25)\right) \quad (8)$$

$$UFL_i[\%] = UFL_{i(25°C)}[\%]\left(1 + 7.21\times10^{-4}(T-25)\right) \quad (9)$$

4 Scaling law and accelerating forest fire cases

Over the last decade, many fire studies implying laws that govern the change in scale - scaling laws - have been discussed by several authors. Hwang and Edwards [28] have shown with numerical simulations that the Froude-scaling law is a good approximation for tunnel fires between model and full-scale. Roh et al. [29] have done experiments on heptane pool fire in tunnels using Froude scaling and have found proportionality between ventilation velocity and heat release rate.

In forest fires literature, we can notice the recent contribution of Pérez et al. [30]. These authors have used a dimensional analysis to find the scaling laws and have studied the effect of changing scale in forest fires experimentation.

To quantify the VOCs concentration at field scale in the canyon geometry proposed by Viegas [1], we will apply the Froude scaling law to our laboratory experiments. The Froude scaling law [31] consists in the preservation between model scale and full scale of the Froude number defined as the ratio of inertia forces to gravitational forces. So, we can write:

$$Fr = \frac{u_{lab}^2}{gL_{lab}} = \frac{u_{can}^2}{gL_{can}} \tag{10}$$

where $g\left[\mathrm{m\cdot s^{-2}}\right]$ is the gravitational acceleration, $u\left[\mathrm{m\cdot s^{-1}}\right]$ and $L\left[\mathrm{m}\right]$ are respectively characteristic values of velocity and length. Subscript "*lab*" refers to laboratory scale and subscript "*can*" to canyon scale. A straightforward calculation from equation (10) gives:

$$\frac{u_{can}}{u_{lab}} = \sqrt{\frac{L_{can}}{L_{lab}}} \tag{11}$$

The concentration of VOCs in the hermetic enclosure and in the canyon at the sampling location and at different temperatures is given by the following relation [3]:

$$C_k\left[\mathrm{g\cdot m^{-3}}\right] = \frac{\dot{m}_k \cdot \Delta t_k}{Q_v \cdot \tau} \tag{12}$$

where $\dot{m}_k\left[\mathrm{kg\cdot s^{-1}}\right]$ and $\Delta t_k\left[\mathrm{s}\right]$ are respectively the mass flow rate of the VOCs emission and the heating time at scale *k* (*can* or *lab*). $Q_v = 150\ \mathrm{mL\cdot min^{-1}}$ is the flow-rate of gases extracted by the pump and $\tau = 10\ \mathrm{min}$ is the sampling time; they are scaling invariants.

Then, we have:

$$C_{can} = C_{lab}\frac{\dot{m}_{can}}{\dot{m}_{lab}}\frac{\Delta t_{can}}{\Delta t_{lab}} \tag{13}$$

The mass flow rate of the VOCs emission can be defined as

$$\dot{m}_k = \rho u_k L_k^2 \tag{14}$$

with $\rho\left[\mathrm{kg}\cdot\mathrm{m}^{-3}\right]$ the density of the VOCs mixture, $u_k\left[\mathrm{m}\cdot\mathrm{s}^{-1}\right]$ their emission velocity and $L_k\left[\mathrm{m}\right]$ a characteristic length deduced from the surface occupied by vegetation at scale k.

Based on the relation (14), the ratio of the mass flow rates and heating times are:

$$\frac{\dot{m}_{can}}{\dot{m}_{lab}} = \left(\frac{L_{can}}{L_{lab}}\right)^{\frac{5}{2}} \tag{15}$$

$$\frac{\Delta t_{can}}{\Delta t_{lab}} = \sqrt{\frac{L_{can}}{L_{lab}}} \tag{16}$$

Using the relations (13), (15) and (16), the VOCs concentration in the canyon can be given by the following expression:

$$C_{can} = C_{lab}\left(\frac{L_{can}}{L_{lab}}\right)^{3} \tag{17}$$

Now, we will apply this formula to three real accidents which occurred in USA (South Canyon accident, 1994), France (Palasca accident, 2000) and in Croatia (Kornati accident, 2007) described respectively by Butler et al. [5], Dold et al. [2], Viegas et al. [7] and Viegas [8]. The canyon characteristic lengths of these accidents are determined using the burned area where the phenomenon happened and the canyon geometry given by Viegas [1] and illustrated in Figure 3. These canyon characteristic lengths L_{can} can be calculated by the following relation:

$$L_{can} = l\cdot\cos\left(\frac{\theta}{2}\right) \tag{18}$$

where θ is the canyon angle and $l = \sqrt{\frac{A_{can}}{2}}$ with A_{can} = 120 ha, 6 ha and 10 ha for the South Canyon, Palasca and Kornati accidents respectively.

By means of the hermetic enclosure results and the equations (17) and (18), we can plot in Figure 4 the evolution of the VOCs mixture concentration involved in these three accidents and the VOCs mixture flammability limits versus temperature.

As we can see in this Figure, the VOCs concentrations under a particular climatic condition (without wind) for the large scale calculations (South Canyon, Palasca and Kornati accidents) are in the flammability domain for 6, 9 and 6 values of temperature respectively. These results show that under certain

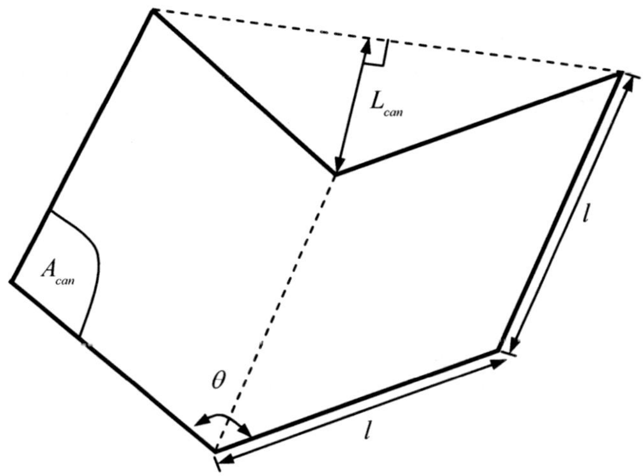

Figure 3: Schematic configuration of canyon shape.

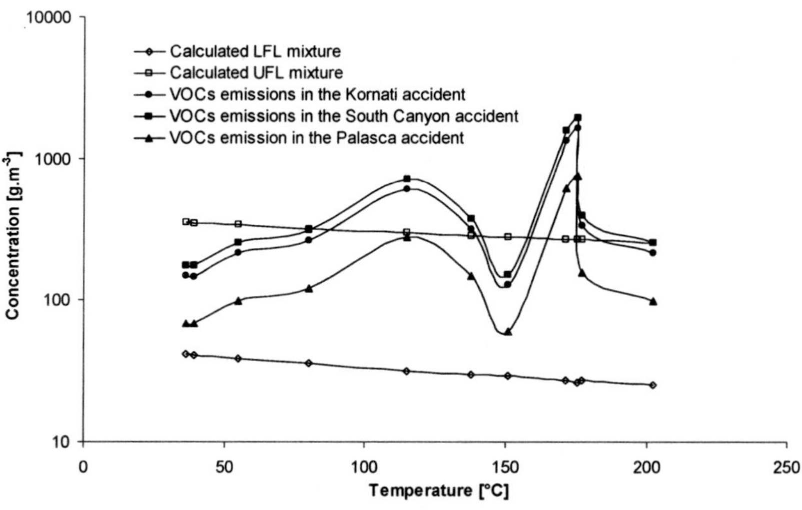

Figure 4: Comparison between VOCs mixture flammability limits and its concentrations in three real accidents.

conditions of vegetation type, climate and topography, it could be possible to have a flammable gas mixture in the vicinity of the vegetation ahead of the fire that could lead to a gaseous flame propagation and to the development of an accelerating fire triggered by this gaseous combustion process for those accidents.

Let us notice that this approach based on two geometrical parameters (the burned area A_{can} and the canyon topology depending on the angle θ) and on the emitted VOCs flammability limits is an explanation for the “lake of fire” observed in the Palasca accident [2] and for the acceleration of the rate of spread in the South Canyon accident. Concerning the Kornati accident, this approach can be considered as an alternative way to the eruptive fire theory developed by Viegas [1] and applied to this accident [7].

5 Conclusion

The present study is conducted on the relation between the VOCs emissions and their potential to cause an accelerating forest fires. It was found with laboratory experiments that *Rosmarinus officinalis* plants heated by external radiant heat flux emit fourteen VOCs, mainly monoterpenes hydrocarbons. The concentration of this gas mixture at canyon scale was calculated from these experimental results by means of Froude scaling law, assuming that there is no flow of VOCs out of the canyon space. The estimation of VOCs concentrations in canyons where three real accidents occurred (South Canyon, Palasca and Kornati), based on these hypothesis, has shown that their values can be in the calculated flammability domain of the gas cloud. Consequently, this phenomenological approach can be used to explain some real cases when those conditions are met. In such conditions, this process can be considered as an alternative explanation to the feedback effect of the convection induced by the fire [1] and to the flow attachment in the direction of the fire spread [14]. It will be interesting to realize in future work experimental studies of VOCs emissions at real scale (in canyons) and to compare them with the obtained results.

References

[1] Viegas, D.X., A Mathematical model for forest fires blow-up. *Combustion Sciences and Technology*, 177, pp. 27–51, 2005.

[2] Dold, J.W., Simeoni, A., Zinoviev, A. & Weber, R., The Palasca fire, September 2000: Eruption or Flashover?. *Recent Forest Fire Related Accidents in Europe*, ed. Viegas, pp. 54–64, 2009.

[3] Chetehouna, K., Barboni, T., Zarguili, I., Leoni, E., Simeoni, A. & Fernandez-Pello, A.C., Investigation on the emission of Volatile Organic Compounds from heated vegetation and their potential to cause an accelerating forest fire. *Combustion Science and Technology*, 181, pp. 1273–1288, 2009.

[4] Rothermel, R.C., Mann Gulch fire: a race that couldn’t be won. *General Technical Report INT-299*, United States Department of Agriculture, Forest Service, Intermountain Research Station, 10 p., 1993.

[5] Butler, B.W., Bartlette, R.A., Bradshaw, L.S., Cohen, J.D., Andrews, P.L., Putnam, T. & Mangan, R.J., Fire behaviour associated with the 1994 South Canyon Fire on Storm King Mountain, Colorado. *Research Paper, RMRS-*

RP-9, Ogden, UT: U.S. Department of Agriculture, Forest Service, Rocky Mountain Research Station, 82 p., 1998.

[6] Furnish, J., Chockie, A., Anderson, L., Connaughton, K., Dash, D., Duran, J., Graham, B., Jackson, G., Kern, T., Lasko, R., Prange, J., Pincha-Tulley, J. & Withlock, C., Thirtymile fire investigation, *Factual Report and Management Evaluation Report*, USDA, Forest Service, 97 p. , 2001.

[7] Viegas, D.X., Stipanicev, D., Ribeiro, L., Pita, L.P. & Rossa, C., The Kornati fire accident – eruptive fire in relatively low fuel load herbaceous fuel conditions. *Modelling, Monitoring and Management of Forest Fires*, ed. J. de la Heras, C.A. Brebbia, D. Viegas & V. Leone, pp. 365–375, 2008.

[8] Viegas, D.X., (ed). *Recent Forest Fire Related Accidents in Europe*, European Commission, Joint Research Centre, Institute for Environment and Sustainability, 2009.

[9] Barboni, T., Cannac, M., Leoni, E. & Chiaramonti, N., The Emission of Biogenic Volatile Organic Compounds implicated in an Eruptive Fire for the Safety of Firefighters. *International Journal of Wildland Fire*, In Press, 2010.

[10] Viegas, D.X., *Cercados pelo Fogo*, Minerva (ed), Coimbra, 2004.

[11] Viegas, D.X., *Cercados pelo Fogo, Parte 2*, Minerva (ed), Coimbra, 2009.

[12] Xanthopoulos, G., Viegas, D.X. & Caballero, D., The fatal fire entrapment of Artemida (Greece). *Recent Forest Fire Related Accidents in Europe*, ed. Viegas, pp. 65–75, 2009.

[13] Viegas, D.X., Parametric study of an eruptive fire behaviour model. *International Journal of Wildland Fire*, 15, pp. 169–177, 2006.

[14] Dold, J.W. & Zinoviev, A., Fire eruption through intensity and spread rate interaction mediated by flow attachment. *Combustion Theory and Modelling*, 13(5), pp. 763–793, 2009.

[15] Ormeño, E., Fernandez ,C., Bousquet-Mélou, A., Greff, S., Morin, E., Robles, C., Vila, B. & Bonin, G., Monoterpene and sesquiterpene emissions of three Mediterranean species through calcareous and siliceous soils in natural conditions. *Atmospheric Environment*, 41, pp. 629–639, 2007.

[16] Barboni, T., Contribution de méthodes de la chimie analytique à l'amélioration de la qualité de fruits et à la détermination de mécanismes (EGE) et de risques d'incendie. *PhD thesis*, Corsica University, France (In French), 2006.

[17] Le Chatelier, H. & Boudouard, O. Sur les limites d'inflammabilité de l'oxyde de carbone. *Comptes Rendus*, 126, pp. 1344–1349, 1898.

[18] Mashuga, C.V. & Crowl, D.A., Derivation of Le Chatelier's Mixing Rule for Flammable Limits. *Process Safety Progress*, 19, pp.112–117, 2000.

[19] Catoire, L. & Naudet, V., Estimation of temperature-dependent lower flammability limit of pure organic compounds in air at atmospheric pressure. *Process safety progress*, 24, pp. 130–137, 2005.

[20] Gharagheizi, F., Quantitative Structure-Property Relationship for Prediction of the Lower Flammability Limit of Pure Compounds. *Energy & Fuels*, 22, pp. 3037–3039, 2008.

[21] Gharagheizi, F., Prediction of upper flammability limit percent of pure compounds from their molecular structures. *Journal of Hazardous Materials,* 167, pp. 507–510, 2009.
[22] Vidal, M., Wong, W., Rogers, W.J. & Mannan, M.S., Evaluation of lower flammability limits of fuel-air-diluent mixtures using calculated adiabatic flame temperatures. *Journal of Hazardous Materials*, 130, pp. 21–27, 2006.
[23] Zhao, F., Rogers, W.J., Mannan, M.S., Calculated flame temperature (CFT) modeling of fuel mixture lower flammability limits. *Journal of Hazardous Materials*, 174(1-3), pp. 416–423, 2010.
[24] High, M.S. & Danner, R.P., Prediction of Upper Flammability Limit by a Group Contribution Method. *Industrial & engineering chemistry research*, 26, pp. 1395–1399, 1987.
[25] Suzuki, T. & Ishida, M., Neural network techniques applied to predict flammability limits of organic compounds. *Fire Materials*, 19, pp. 179–189, 1995.
[26] Arnaldos, J., Casal, J. & Planas-Cuchi, E., Prediction of Flammability limits at reduced pressures. *Chemical Engineering Science*, 56, pp. 3829–3843, 2001.
[27] Zabetakis, M.G., Flammability characteristics of combustible gases and vapors. *Bureau of Mines Bulletin 627*, Washington, DC, 129 p., 1965.
[28] Hwang, C.C. & Edwards, J.C., The critical ventilation velocity in tunnel fires-a computer simulation. *Fire Safety Journal*, 40, pp. 213–244, 2005.
[29] Roh, J.S., Yang, S.S., Ryou, H.S., Yoon, M.O. & Jeong, Y.T., An experimental study on the effect of ventilation velocity on burning rate in tunnel fires-heptane pool fire case. *Building and Environment*, 43, pp.1225–1231, 2008.
[30] Pérez, Y., Àgueda, A., Pastor, E. & Planas, E., Study of the effect of changing scale in forest fires experimentation by means of dimensional analysis. *Forest Ecology and Management*, 234, pp. S113, 2006.
[31] Quintiere, J.G., Scaling applications in fire research. *Fire Safety Journal*, 15, pp. 3–29, 1989.

Forest fires, risk and control

H. Azari
University College of Borås, School of Engineering, Sweden

Abstract

Fire is known as a primitive and amazing effect of nature and for humans it can be a friend, source of light and heat. Human intervention and ignoring nature's rules can lead to an out of control forest fire, changing its face into an uncontrollable force. Through the thesis work, more than 450 articles, scientific papers, and web-pages have been studied and reviewed to find points and comments that might help to reduce the fire occurrence probabilities, fire aggression, and damage to people, the environment, and the economy.

Ten years of work experience in industrial safety and fire fighting has been applied to new findings in the field of forest fires. There are many points and comments reviewed in the thesis document and here through this paper some specific results are been discussed. The introduction of the concept of *layer protection* to this study has been attempted, in which *preventive* and *remedial measures* are the two main schemes for reducing unwanted incidents and their probable consequences. To prevent the risk, deviation should be indicated, while before that it is necessary to set main roles or standards. In this situation, failure-elimination is the first step toward layer protection, and failure-control would be placed in the next layer. To reduce the consequences of failure, the last layer should be carefully positioned. According to findings during the survey, a suggestive concept has been recommended according to zoning theory, which fragments high-fire-risk forests to limit the fire aggression. A safety swath network is recommended based on this theory, with flexible width suggested of between 30 and 70 meters. This safety network could be augmented by a specified grid of a water mist network with the ability of humidifying approximately 100 meters of another zone with the application of a mobile sprinkler system. For populated areas, such as camping sites, a safety boundary has been proposed to isolate these risky areas from the forest. There are always activities that can stop autonomous disasters or reduce their consequences; our task is to find these activities and adhere to them. The aim of the study is to

WIT Transactions on Ecology and the Environment, Vol 137, © 2010 WIT Press
www.witpress.com, ISSN 1743-3541 (on-line)
doi:10.2495/FIVA100211

condense existing literature in a collection and make new suggestions to improve forest treatment and make that knowledge accessible for students, researchers, and forest management systems. Reducing forest fire consequences would be our final goal in a responsive collaboration toward a safer future.

Keywords: layer protection, preventive and remedial measures, passive and active protection, zoning theory, safety swath network, safety mound network, safety boundary rand, water mist network.

1 Introduction

Flood, forest fire, landslide, storm, ozone depletion, air pollution, waste and garbage problems, and industrial accidents are some examples of natural or industrial disasters that almost every territory is subjected to. We might not be sure about the causes of these flourishing catastrophes, but their consequences are unbearable. Human interventions through climate change, fuel accumulation, ignition, and abandoned agricultural lands could be some of the reasons for an unusual number of wildfires. Meanwhile, intensive forest fires could have an amplification effect on themselves through adding their CO_2 emission to the human share, inducing global warming and climate change.

Forest fire has damaged *over 10 million hectares (ha)* of Southern European forests in the two last decades [5]. Every year, in the Mediterranean Basin, wildfires burn down 700,000-1,000,000 ha of wildlands [6]. Throughout Europe this figure is about 500,000 ha of forested areas caused by 50,000 fires [7]. In Spain, during the last 40 years, forest and bush fires have burned down around 6,400,000 ha of green-lands with economic damage of around 7.5 billion Euros [39]. Forest fire impact in Portugal is also significant – over 15 years (1990-2005) nearly 25% of the countries forest area has been burned, which is about 2.2 million ha [4].

In Sweden, forest fires have not been too extensive and they can be put out in less than a few hours. Apart from precipitation, access roads are very important in this regard and in Sweden the average span to a navigable road is between 400 and 500 m [16, 17]. The history of forest fire frequencies in northern Sweden show intervals of between 60 and 80 years [9, 13], while the average fire intervals in the boreal forests is between 50 and 200 years [17, 24]. According to a study published in 1998, *mitigation of forest carbon emissions* in some European countries can *compensate* 30-35% of their carbon emissions, while this figure in Sweden proved to be about 60% [36].

All forest fires are bad and should be put out in a minimum time was fire policy in the US and after the death of 34 fire fighters during the fire season in 1994, a concern about fire policy was raised in order to find better treatments for forest fires [41].

Canada owns more than 10% of the world's forest [31] and the forest industry contributes $33 billion to the gross domestic products and directly employs 376,000 people [10]. Every year in Canada, almost 10,000 fires burn about *2.5-2.8 million ha* of wildland areas, which are caused by human activity and lightning [20, 23, 31]. During the fire season, forest fire damage to public and private resources could easily reach hundreds of million dollars, while fire suppression expenses easily hit $1 billion [10]. A burned area for *one-wildfire* in the range of 100,000 ha is not uncommon and 1 million ha has been recorded as well. Over 20 years (1980-2000) the mean burned area in Canada has increased dramatically [34]. Over the next century, considering climate change, wildfire severity could substantially increase so that by 2040, in Canada, annually the burned area could be doubled and wildfire problems would go beyond our present and future capacity. On the other hand, forest fire exclusion could accumulate fuel and encourage insect infestation with the consequence of dead fuel accumulation and older non-healthy forests, which are good choices for wildfires [10].

Siberian forests are protected by one amazing natural effect, *Permafrost*, which preserves ice from rain water in the soil. Recently, increasing forest fires have raised absorption of solar radiation into the ground, disturbing the permafrost and changing forests to swamps or grasslands in an irreversible way [22]. The African continent owns two-thirds of the Earth's savannas and burning savannas in Africa play an important role in CO_2 production [26].

Unwanted wildfires could be a threat to public health, property, forest services and non-market values, while forest fires would improve the biodiversity, health and productivity of the forest ecosystem, considering the biological, ecological and physical characteristics of wildland fires that could be performed through natural or prescribed fires. Risk identification and fire management, including public education, preparedness, response, mitigation, recovery, and land management by fire, are essential activities for improving forest health [12], in which risk could be defined as the likelihood of an unwanted and unpleasant incident and its probable consequences [3, 27].

Bio-fuel in the forest could be a source of energy for humans, but if forest management systems do not take additional fuel at the right time, fires would take it, resulting in unbearable consequences. There are important issues in forests, which impose adverse impacts on human life and the environment. Meanwhile, some of them could be availed for human use and somehow preclude damage to the environment. Discrimination and confusion are two essential facts that should be distinguished among different causes and effects of forest fires, so that clear intuition through different parameters in the forest emerges. For example, the "All forest fires are bad" concept has been changed to new and more realistic concepts.

2 Materials and methods

Through the review study, more than ten years of experience in industrial safety and fire fighting is combined with new findings in the forest fire field. Articles, scientific papers, and web-pages have been studied and reviewed to find points and comments that might help to reduce fire occurrence probabilities, fire aggression, and damage to people, the environment, and the economy. A HSE (Health, Safety, Environment) industrial plan has played an important role in connecting industrial concepts of safety to forest fire facts (Fig. 1). In the first step, it has attempted to extract results from literature to be used in the concept of layer protection, which is adapted from the industrial safety and fire fighting context. This concept is introduced to the study by means of *preventive* and *remedial measures*, which are two main schemes for reducing unwanted incidents and their probable consequences.

3 Results: a suggestive concept

Among existing ways of controlling wildfires, forest management may consider conventional fuel treatment, including thinning, branch pruning, bush and slash removal and some extra precautionary measures, such as fuel breaks, access roads, introducing high fire resistance plants and keeping a low surface fuel load, utilizing domestic animals [8]. In many countries, there are effective activities for forest fuel management and risk mitigation, but it is difficult to do this for all territories and countries. However, forest fires should be kept in a rational, moderate, and controllable criterion. Systematic suppression, fuel accumulation throughout years without regular fuel reduction, and above all drought, has made forests vulnerable to fire. As a result, high fuel density forests provide suitable conditions for intensive fires [3]. Different forests should have their own policies, depending on the management ability of the fuel control. It would also be possible to try new suggestions that limit the fire progress in a specific

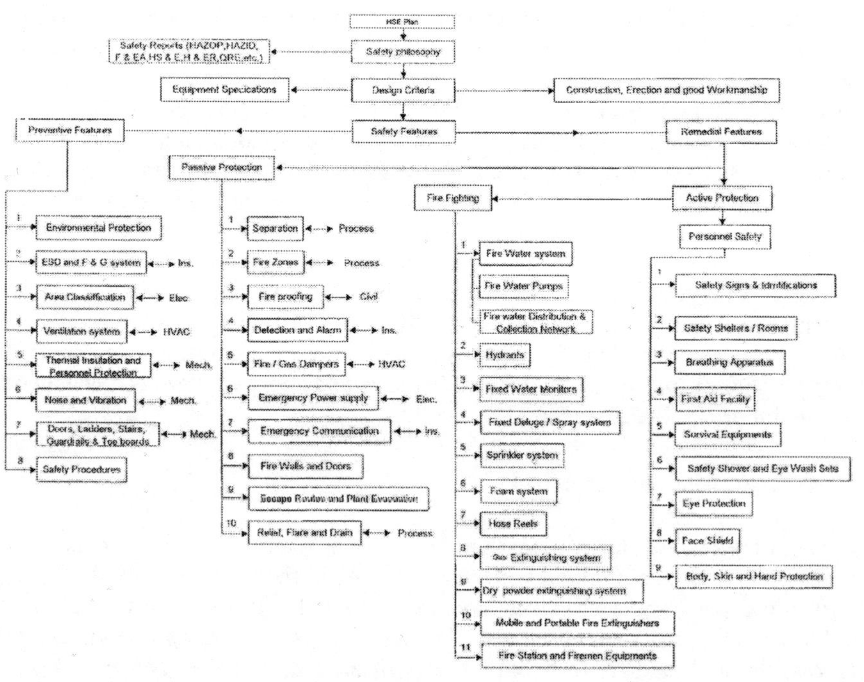

Figure 1: HSE industrial plan as the study base.

manner. According to the above mentioned HSE plan, fire protection could be divided into two main sections: passive and active protection. Separation and fire zoning are pivotal aspects in passive fire protection, which could be utilized in forest fire protection. Accordingly, some suggestions will be discussed according to these aspects and in future works they might be developed to theoretical and operational activities. For many months and years, it would not be rational to wait for fire, despite having equipment, which most probably is not adequate when wildfire attacks.

3.1 Zoning theory

There could be a simple and unique solution for many kinds of forests against wildfires, which is the application of *Zoning Theory*. This theory would divide the forest into different zone areas. We might not be able to specify an area size description for all territories, but each country could decide the size of the zone area according to its *territory*, *fire risk*, and *equipment*. Forest fires could be divided into *easy*, *difficult*, and *impossible to be tackled* by the fire management system. Accordingly, fire consequences through this kind of management might be tolerable, laborious, and insupportable, respectively. Depending on fire types, the area size might be small or extensive and of any shape or topographic configuration. For instance, a forest with an *easy-to-tackle* condition could have a larger zone area than another forest with *difficult* conditions.

To be more understandable, an area of 400 ha – any geometric shape driven from the territory, or 2.0 Km by 2.0 Km – could be considered for a territory with medium fire risk as a base zone area, which is less than the considered area for the "zone" in other documents [14, 42]. Among different areas, there might be zones with *high* or *low* fire risk and in turn, zone-areas would be *smaller* or *larger*, respectively. The zone-area could be *very limited* in a region with *arson risk*. Within the zone-area fire would be suppressed or *the forest itself* would be able to limit the fire in one or maximum two zones. If the forest has reached the time for prescribed burning – prescribed natural fire – or there are limitations to our extinguishing ability, or the fire goes wild, it would be allowed to burn, since it cannot burn more than itself – our battery limit.

Through this zoning theory, there is the possibility to order areas based on their fire season. For example, if we decide to put a 100 year interval for fire, *zone patches would be scattered and fire occurrences in one zone would be confronted with non-uniform forest textures* and this could be one step toward controlling the fire. For example, we would have a 20 year interval between patches that start from 0 and continue to 20, 40, 60, 80, and 100 year zone-area patches.

This could be a basis for the least infrastructure for every forested territory that has suffered from fire. In the Cape of Creus, NE of Catalonia, Spain, there are areas that have been burned five times just during the 21 years from 1975 to 1995 [33]. These areas could be a good choice for applying zoning theory. *There would also be special a dynamic fire-break between zones, named the Safety Swath.*

3.2 Safety swath network

Roads and access-ways are fundamental elements in limiting forest fire through their double acting phenomena, as a fire break and provider of rapid intervention. The safety swath is not a road but a developed space and the master key in *Zoning Theory*, which should protect and isolate zones as a fire break. It is called a dynamic because it could have dynamic specifications, such as non-fixed wideness or non-fixed moisture conditions (Fig. 2).

Performing this extensive infrastructure would take time and expenditure and it is better to be accomplished gradually – from high fire risk to low fire risk areas. It should be noted that if there are limitations to applying this theory, only high risk areas could be considered and all forest areas might not need this zoning system.

The cost of performing the safety swath could be its design, implementation, enforcement, and cleaning after one or two years. Special cutting could be done in the case of a fire incident, which is obviously an easy job having this infrastructure. In the case of outraging wildfires there would be a possibility of starting a back-fire from the other side of the zone and toward the wind direction to widen the safety swath.

This theory could also be applied in WUI and major-value-at-risk points. Apart from defensible space that protects every house, there would be a safety swath that isolates one zone from another. The area of the zone should be determined according to the fire risk and terrain conditions.

3.3 Safety swath specification

First and foremost the width of this safety band should be indicated. According to investigation through available documents, it should not be so narrow so as to act as a fire-ladder and not so wide that it cannot be handled. In the design condition, we speak carefully about 500-600 m for the width of a firebreak, but when it comes to a fire situation *we look for just a 3 meter road*. In practice there are 8-12 m firebreaks, including natural creeks, that functioning in a network [2]. It is suggested that the optimum distance of the forest from a house is about 50 m [11, 19, 30, 32, 40], which could be an applicable figure after some considerations.

Figure 2: Safety swath network augmented by water mist network.

If there are economical problems in providing a 50 m interval between zone areas, considering the prevailing-wind during the fire season, there would be a possibility of reducing the width of this safety band. To do this, the angle between the prevailing-wind during the fire season and the safety swath middle line (α) should be established. The following formula would be a good start for considering limitations:

$$\text{Width (m)} = 30 + 20 \operatorname{Sin}(\alpha) \qquad (\text{slope} < 5°)$$

As can be seen from this formula, 50 m has been divided into a base distance of 30 m plus a maximum 20 m variable.

According to the results detailed in the wind and slope section in the thesis [35], every 10° slope change, up to 15°, would tilt the fire angle approximately 15°. Considering the maximum 40 m for flame length, there would be around 10 m flame displacement, D = 40 tg (15°) = 10.7 m. To compensate for this change and to increase the safety margin in maximum 15° slopes, the width of this band might be increased by another 20 m. Width (m) = 50 + 20 (5° < slope < 15°). For slopes more than 15°, further research is required. All of these assumptions have been suggested according to existing research results and require more detailed investigations. Fuel through this swath could be grass that would be treated by means of systematic grazing or machines – at least once in two years. Around particular zone-areas could be cleaned during fire incidents, which would not be a major task. *Slopes that are facing toward the prevailing-wind should be considered as more vulnerable areas and augmented with special features.*

3.4 Safety mound network: the second choice

Since the forest terrain could somehow disrupt the fireline intensity [1, 15, 18, 25, 29, 37, 38] in high fire risk areas, there would be the possibility of building a network of artificial *mounds* – 15 to 20 m high – along with specific considerations. Trees should not grow on the *mound* and it could be protected by the water-mist network (Fig. 3). In this way, *the mound would absorb the radiation energy of wildfire* and present a good opportunity to put out the wildfire. On the other hand, at the other side of the mound, *wind speed would be restricted* and the fire has less tendency to come down in the location of the fire fighters who are there to put it out. Washing by rain and construction expenses are two important problems with this choice.

Figure 3: Safety mound network augmented with water mist system.

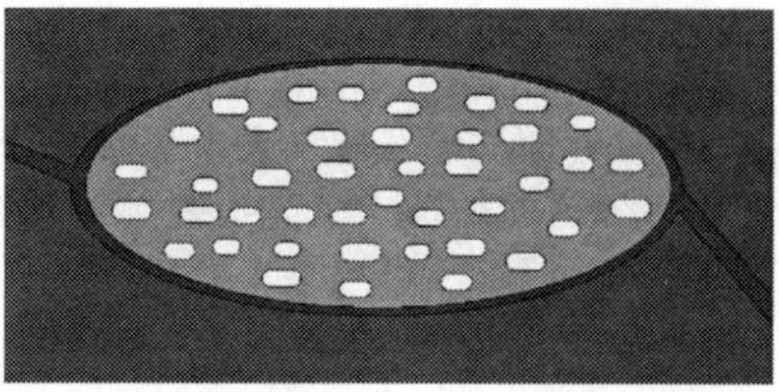

Figure 4: A schematic diagram for SBR, applicable to a camping area.

3.5 Safety boundary rand (SBR)

Forest crossing roads or railways, camping areas, and other civil centers should be equipped with SBR, which could start at least 20 m from the edge. This rand should be protected and cleaned from easily flammable shrubs, teasels, and bushes. To protect camping and recreation centers, it would be a good idea to place these important areas in natural or artificial islands as safety boundary rands. Through this scheming, forests would, to some extent, be protected and these infrastructures would also be protected from fire. In addition, these surrounding rivers or water rings would bring more a pleasant environment to camping and recreation centers – e.g. fishing activities (Fig. 4).

3.6 Water mist network

After evaluating fire risk through all territories, all high fire risk zones could be considered in the water mist network concept. This network applies along with the safety swath or the safety mound to add more support to our fire break during intense fires (Fig. 5). Most fire prone areas could be protected by this *water mist protection*, which would humidify at least 100 m of another zone.

There is a suggestion for applying a mobile sprinkler system as a safety band to increase the resistance against fire – particularly in water restricted areas. To fight fire directly, there must be at least 3-5 lit/m^2 water available, while *applying the safety sprinkler system would require only 1 lit/m^2* [21, 28]. Considering the protection of one side of the zone area with this concept, the protected area should be considered as 2000 m (as the zone-area side) by 100 m (as depth of safety band) = 200,000 m^2. Accordingly, the amount of required water would be 1 lit/m^2 X 200,000 m^2 = 200 m^3. The progress of fire in grasslands is relatively high and about 1100 m/h [43], which could be considered as a basis for calculation. As a result, the time for humidification could be considered as one hour. This means that 200 m^3/h is required to protect 2000 m on one side of the zone area through portable sprinklers.

After the management system accepts the water mist protection, this network would be provided in high fire danger areas according to the mentioned hypothesis (Fig. 5). To bring this concept under economical consideration and according to water requirements (200 m^3/h), an underground network of 4 inch plastic pipe could be considered. Presuming this pipe size, the maximum water velocity in the pipe would be 3.6 m/s which is an acceptable velocity.

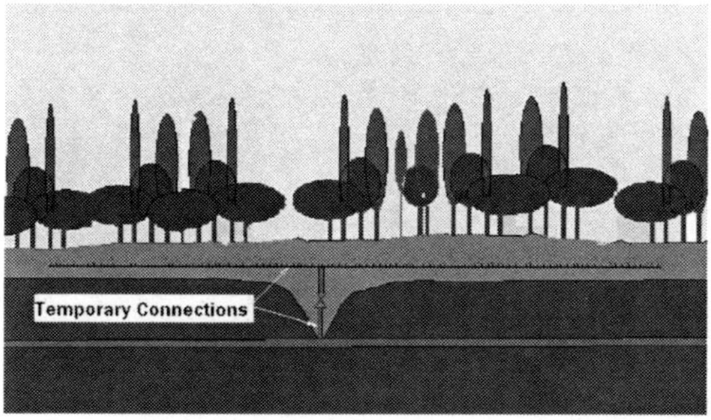

Figure 5: A schematic diagram for the water mist system.

The installation of a portable water mist sprinkler on an existing underground network would not be a time consuming job. Assuming 100 m for each portable water mist package, 20 fast-connections should be made on the main underground plastic pipe. Another quick alternative would be a 2000 m package, which should be connected to the network at two end points.

Providing quick access to fire water for fire trucks could be another advantage of this water network. In addition, distributed fire-water storage basins, particularly in elevated areas for hastening the fire fighting process, are effective forest fire repressors as fast providers of fire-water for fire trucks, air-tankers and the possible network system. Further experimental tests and calculations are required to establish final conditions based on this theory.

3.7 Soil improvement

Preserving water in the soil like *Permafrost* could be one effective way of protecting forests from fire through *adding high water absorbent gels* to the soil after evaluating the economic and possible side effects of this method. Soil water content is an important parameter in starting forest fires and there is room for research in soil improvement, such as *high water absorbent gels*, which could be tested and introduced to thin-soil areas or soils with low water tenability to improve water composition in the soil.

4 Discussion

In industry, usually the mentioned safety configuration plan has been used (Fig. 1). In the case of forest fire many of those fire controlling aspects could be applied. Summarizing all the mentioned steps and activities, the following *forest rescue plan* as a block diagram (Fig. 6) has been prepared based on study results and prior experiences. It starts with the preparation of an environmental and safety philosophy including, safety reports. The preparation of design criteria would be the next step, including forest and job specifications. In the case of forest fire, passive protection is combined with preventive measures under the safety feature box.

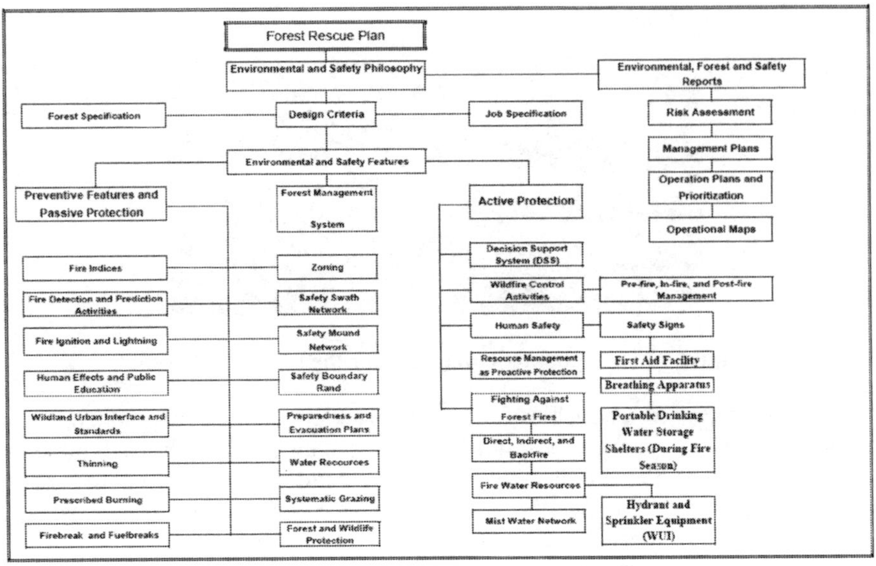

Figure 6: Forest rescue plan.

Passive protection could be part of forest management or activities that are done to reduce the risks of a probable fire, such as zoning, instant fire alarms, human and environmental protection. Preventive measures contain a wide range of actions, including people training and forest management.

In active protection human protection should warrant the most attention and afterwards there would be fighting against fire, utilizing different kinds of equipment to overcome the fire. A decision support system, resource management, fighting methods, and wildfire control activities are placed under the active protection box.

There are many points and comments in reviewing thesis documents and here through this paper some specific results have been discussed. There is room for investigating new and more-effective concepts, despite many improvements in managing forest fires. Among existing ways of controlling forest fires, there could be the possibility of trying new suggestions that limit the fire progress in a specific manner. Zoning theory is a simple, unique, and basic solution for many sorts of forests against wildfire. This theory would divide the forest into different zone areas. The zone-area could be *very limited* in a region with *arson risk.* Through this zoning theory, there is the possibility of ordering areas based on their fire seasons to make a non-uniform forest texture and thereby reducing wildfire aggressiveness.

Acknowledgement

The work is based on a master thesis with the supervision of Professor Håkan Torstensson.

References

[1] Albini F.A., Behavior: A User's Manual. USDA Forest Service, Intermountain Forest and Range Experiment Station, Ogden, UT, 68 pp. 1976.

[2] Hoare Peter, A process for community and government cooperation to reduce the forest fire and smoke problem in Thailand, Agriculture, Ecosystems and Environment 104 (2004) 35–46.

[3] Roloff Gary J., Mealey Stephen P, Clay Christopher, Barry Jeff, Yanish Curt, Neuenschwander Leon, A process for modeling short and long-term risk in the southern Oregon Cascades, Forest Ecology and Management 211(2005)166–190.

[4] Carreiras Joao M.B., Pereira Jose′ M.C., An inductive fire risk map for Portugal, Forest Ecology and Management 234S (2006) S56.

[5] Alonso-Betanzos Amparo, Fontenla-Romero Oscar, Guijarro-Berdinas Bertha, Hernandez-Pereira Elena, Andrade Maria Inmaculada Paz, Jimenez Eulogio, Soto Jose Luis Legido, Carballas Tarsy, An intelligent system for forest fire risk prediction and fire fighting management in Galicia, Expert Systems with Applications 25 (2003) 545–554.

[6] P. Fernandes, H. Botelho, Analysis of the prescribed burning practice in the pine forest of northwestern Portugal, Journal of Environmental Management 70 (2004) 15–26.

[7] Keramitsoglou Iphigenia, Kiranoudis Chris T., Sarimveis Haralambos, Sifakis Nicolaos, A Multidisciplinary Decision Support System for Forest Fire Crisis Management, Environmental Management Vol. 33, No. 2, pp. 212–225.

[8] Kaloudis Spiros, Tocatlidoub Athena, Lorentzos Nikos A., Sideridis Alexander B., Karteris Michael, Assessing Wildfire Destruction Danger: a Decision Support System Incorporating Uncertainty, Ecological Modelling 181 (2005) 25–38.

[9] Pahlén Tina, Att restaurera forna tiders beståndsstruktur. Ett exempel från Jämtgaveln, SLU, Institutionen för skoglig vegetationsekologi, Examensarbete i skoglig vegetationsekologi, juni 2000.

[10] A report to the Canadian Council of Forest Ministers, prepared by the Canadian Wildland Fire Strategy, Assistant Deputy Ministers Task Group, Canadian Council of Forest Ministers, 2005, Canadian wildland fire strategy: A vision for an innovative and integrated approach to management risks, Catalogue No. Fo134-1/2005E-PDF, ISBN 0-662-42195-7.

[11] Howard R.A., North D.W., Offensend F.L., Smart C.N., 1973. Decision Analysis of Fire Protection Strategy for the Santa Monica Mountains: an Initial Assessment. Stanford Research Institute, Menlo Park, CA.

[12] Canadian Council of forest ministers, Declaration of Canadian Wildland Fire Strategy, Saskatoon, Saskatchewan, 2005.

[13] Linder P. 1988. Jämtgaveln - En studie av brandhistorik, kulturpåverkan och urskogsvärden i ett mellannorrländskt skogsområde. Rapport 1988:3, Länsstyrelsen i Västernorrlands Län. Härnösand, Sweden.

[14] Simard A., 1991. Fire severity, changing scales, and how things hang together. Int. J. Wildland Fire 1 (1), 23–34.

[15] Rothermel, R.C, 1983. Forest and Range Fires. Gen. Tech. Rep. INT-143. USDA, forest service, intermountain forest and range exp. sm. ogden, UT, 161 pp.

[16] Granström A. (1998): “Forest fire and fire management in Sweden”. International Forest Fires News, no 18.

[17] López Jerónimo, Forest fires and fire management in Sweden; a comparison with Spain, Examensarbete Nr. 24, 2003, Institutionen för skogens produkter och marknader, SLU, ISSN 1651-4467.

[18] Rothermel, R.C., Wilson, Jr., R.A., Morris, G.A. and Sackett, S.S., 1986. Fuels: Input to the BEHAVE Fire Prediction System. Res. Pap. INT-359. USDA Forest Service, Intermountain Res. Stn. Ogden, UT. 61 pp.

[19] Collins Timothy W., Households, forests, and fire hazard vulnerability in the American West: A case study of a California community, Environmental Hazards 6 (2005) 23–37.

[20] Lee B.S., Alexander M.E., Hawkes B.C., Lynham T.J., Stocks B.J., Englefield P., Information systems in support of wildland fire management decision making in Canada, Computers and Electronics in Agriculture 37 (2002) 185-198.

[21] Linkewich A., 1972, Air attack on forest fires, DW. Friesen, Calgary.

[22] Chikahisa T., Anzai H., Hishinuma Y., Kudo K., Modeling and evaluating the effect of forest fire control on the CO2 cycle in Siberia, Energy 30 (2005) 2261–2274.

[23] Groot William J. de, Modeling Canadian wildland fire carbon emissions with the Boreal Fire Effects (BORFIRE) model, Forest Ecology and Management 234S (2006) S224.

[24] Schimmel J. (1993): “On fire: fire behaviour, fuel succession and vegetation response to fire in the Swedish boreal forest”. Umeå, Sveriges Lantbruksuniversitet: Reprocentralen, SLU.

[25] Schroeder, M.J. and Buck, C.C., 1970.for Application of Meteorological Information to Forest Fire Control Operations. USDA Forest Service, Agricultural Handbook 360. US Government Printing Office, Washington, DC, 229 pp.

[26] Beaudoin Laurent, Gademer Antoine, Amir Ahmed, Avanthey Loica, Germain Vincent, Pocheau Alexandre, Near real time detection of hot spots on Meteosat Second Generation images : from forest fires to volcanic eruptions, 1-4244-1212-9/07/$25.00 ©2007 IEEE.

[27] Fairbrother Anne, Turnley Jessica G., Predicting risks of uncharacteristic wildfires: Application of the risk assessment process, Forest Ecology and Management 211 (2005) 28–35.

[28] Rickard Hansen, “Skogbrandsläckning”, Utgivningsår 2003, Räddningsverket, Sweden.

[29] Rothermel. R.C., 1972. Spread in Wildland Fuels. Res. Pap. INT-115. USDA Forest Service, Intermountain Forest and Range Exp. Stn., Ogden, UT, 40 pp.

[30] Foote, E.I.D., 1994. Structure survival on the 1990 Santa Barbara ''Paint'' fire: a retrospective study of urban–wildland interface fire hazard mitigation factors. M.S. Thesis, University of California, Berkeley.
[31] Li Z., Chilar J., Remote Sensing of Canadian Forest Fires: Hotspots, Burned Area, and Smoke Plumes, 0-7803-5207-6/99810.00, 1999 IEEE.
[32] Cohen Jack, Johnson Nan, Walther Lincoln, Saving Homes from Wildfires: Regulating the Home Ignition Zone, Zoning News, American Planning Association, APA, May 2001.
[33] Diaz-Delgado Ricardo, Pons Xavier, Spatial patterns of forest fires in Catalonia (NE of Spain) along the period 1975-1995, Analysis of vegetation recovery after fire, Forest Ecology and Management 147 (2001) 67-74.
[34] Podur Justin, Martell David L., Knight Keith, Statistical quality control analysis of forest fire activity in Canada, Can. J. For. Res. 32: 195–205 (2002).
[35] Morandini F., Santoni P.A., Balbi J.H., The contribution of radiant heat transfer to laboratory-scale fire spread under the in influences of wind and slope, Fire Safety Journal 36 (2001) 519-543.
[36] Nabuurs G. J., Paivinen R., Sikkema R., Mohren G. M. J., The role of European forest in the global carbon cycle-A review, Biomass and Bioenergy Vol. 13, No. 6, pp. 345-358, 1997.
[37] Kushla John D., Ripple William J., The role of terrain in a fire mosaic of a temperate coniferous forest, Forest Ecology and Management 9.5 (1997) 97- 107.
[38] Albini, F.A., Latham, D.J. and Baughman, R.G., 1982.Upslope Convective Windspeeds for Predicting Wildland Fire Behavior. Res. Paper NT-257. USDA Forest Service, Intermountain Forest and Range Exp. Stn., Ogden, UT, 19 pp.
[39] Nunez-Regueira Lisardo, Rodriguez-Anon Jose A., Proupn-Castineiras Jorge, Using calorimetry for determining the risk indices to prevent and fight forest fires, Thermochimica Acta 422 (2004) 81–87.
[40] Dombeck, M.P., Williams, J.E., Wood, C.A., 2004. Wildfire policy and public lands: integrating scientific understanding with social concerns across landscapes. Conservation Biology 18 (4), 883–889.
[41] http://www.csfs.colostate.edu/, Wildfire Policy in Transition, Colorado State Forest Service, Colorado State University (Consulted Nov. 2008).
[42] Hardy Colin C., Wildland fire hazard and risk: Problems, definitions, and context, Forest Ecology and Management 211 (2005) 73–82.
[43] Garnica J. G. F., Gonzalez D. A. M., Solorio J. D. B., Forest fire behaviour in prescribed burns under different environmental conditions in Mexico, Forest Ecology and Management 234S (2006) S131.

Spatial distribution of human-caused forest fires in Galicia (NW Spain)

M. L. Chas-Amil[1], J. Touza[2] & J. P. Prestemon[3]
[1]Department of Quantitative Economics, University of Santiago de Compostela, Spain
[2]Department of Applied Economics, University of Vigo, Spain
[3]Southern Research Station, USDA Forest Service, Research Triangle Park, USA

Abstract

It is crucial for fire prevention policies to assess the spatial patterns of human-started fires and their relationship with geographical and socioeconomic aspects. This study uses fire reports for the period 1988-2006 in Galicia, Spain, to analyze the spatial distribution of human-induced fire risk attending to causes and underlying motivations associated with fire ignitions. Our results show that there are four distinctive types of municipalities in this region according to the incidence of intentional agricultural-livestock fires, pyromaniacal behavior, negligence, and unknown causes. They highlight that study of the spatial properties of the human causes and motivations of forest fire activity can provide valuable information for detecting the presence of non-random clusters of fires of various causes in particular locations, where fire management planning should be evaluated more in depth.
Keywords: forest fires, intentionality, negligence, Galicia.

1 Introduction

Forest fires can have devastating effects on forest ecosystems. The major impacts include loss of wildlife habitat; destruction of the vegetation and biomass; worsening of soil productivity and erosion; heating of water and increased sedimentation, with significant effects on aquatic organisms; and damaging atmospheric emissions including smoke and carbon. Forest fires may be the result of natural phenomena (i.e., mainly lightning), human negligence,

WIT Transactions on Ecology and the Environment, Vol 137, © 2010 WIT Press
www.witpress.com, ISSN 1743-3541 (on-line)
doi:10.2495/FIVA100221

accidents (often lumped into a category called "accidental"), and human intentional behavior (i.e., those ignited with malice or set with illegal intent, typically labeled "arson"). The South of Europe - Portugal, Spain, France, Italy and Greece - has been seriously affected both economically and environmentally by these kinds of human-ignited wildfires. In 2008 alone, these five countries registered more than 36,000 wildfires of all causes, burning roughly 160,000 ha, and approximately 70% of these wildfires correspond to Spain and Portugal (European Commission [1]). Because the vast majority of wildfires in the region are not natural, most of the fire activity in Southern Europe is connected to human starts. In other words, fires can be understood as a human mediated event process (Martín [2]). For example, Spanish data show that 4% of the forest fires are caused by lighting; 36% are caused by human-negligence or accidents; and 42% are human intentional actions. In Portugal, intentional fires represent 42% and accidents or negligence represent 58.6% of the human-caused fires. The high rate of human-caused fires in the region imply that positive actions taken by wildland managers to prevent such starts through education could be of great benefit. With enhanced understanding of the underlying drivers or correlates of human-caused fires, the effectiveness of wildfire prevention programs could be further enhanced, allowing for targeted interventions. However, a prerequisite of such enhanced programs is expanded research and development.

Worldwide, agricultural activities have made use of forest firesetting, as a common practice for centuries (Gamst [3], Doolittle and Lightsey [4], Vélez [5], Kuhlken [6]). In addition, social motivations such as revenge against a landowner, as an act of protest, as an attempt to cover up another crime, or as vandalism have been pointed to as underlying drivers of illegal firesetting. However, this human component of the risk of fire occurrence is quite complex to model, and it is therefore rarely included in the statistical analysis of fire danger. There are significant differences in the temporal and spatial scale of biophysical (e.g., weather, fuel conditions) and socio-economic processes (e.g., ownership, labor market conditions) that influence the risk of fire occurrences. In addition, a large share of human-caused fires is intentional, incorporating complex individual motivations. The multifaceted nature of these fires has constrained the number of quantitative studies of them. This paucity of studies stands in the face of evidence of the potential negative economic and ecological effects of these fires in both southern Europe and elsewhere and the concerns of both forest managers and law enforcement organizations that seek to limit such firesetting.

In the US, arson fires account for up to 80% of all fire ignitions in some states. Although there are some negative trends in the rates of such fires, these fires account for a steady or rising share of all area burned in many locations, including US national forests (Prestemon and Butry [7]). Arsonist firesetting has been documented extensively in the United States (Bertrand et al. [8], Gamst [3], Bertrand and Baird [9], Doolittle and Lightsey [4], Kuhlken [6], Maingi and Henry [10]). Recent research has shown that these fires are clustered in time, i.e., they usually occur in concentrated periods (Prestemon and Butry [11]). They also cluster in space, and in a space-time dimension (Doolittle and Lightsey [4],

Butry and Prestemon [12]). In this line of research, Genton et al. [13] show that arson and lightning are the leading causes of wildfires in Florida and that fires ignited by trains, lightning, and arsonists are spatially more clustered than ignitions by other accidental causes.

In Europe, studies which focus on intentional fires using long-term series are scarce. In the particular case of Spain, Martínez and Chuvieco [14] study the spatial distribution patterns at the municipality level of fire properties (e.g. mean burned area, mean time of extinction, forest ownership), including if the fires were intentional, negligent or of an unknown cause. They are able to classify Spanish municipalities as falling into one of four types, according to these general forest fires properties. Other studies have focused on the differences in the spatial or temporal behavior between natural and human-caused wildfires (Vázquez and Moreno [15], Amatulli et al. [16], Benavent-Corai et al. [17]). For example, Amatulli et al. [16] find that fires due to human causes in Aragón are more spatially diffuse than lightning, observing many small hotspot areas. Benavent-Corai et al. [17] detect that human impact has important implication decreasing the inter-event interval and increasing the sparking frequency on forest-fire modeling.

This research analyzes the spatial distribution of causes and motivations of intentional fires that characterised wildfires in Galicia (Northwest Spanish region). Following Martinez and Chuvieco [14], it proposes a zonification of the Galician territory, clustering municipalities of homogeneous fire properties. However, we focus on the causes of fires and incorporate the underlying motivations behind intentional forest fires. The resulting typology of Galician municipalities according to the occurrences of different fires' causes and motivations can provide useful information that could enhance the effectiveness and efficiency of prevention activities. The paper is organised as follows. In the second section, we develop a cluster analysis of forest fire causes and underlying motivations. Following this, we introduce the database and the causality variables. Subsequently, we highlight the main causes and motivations of firesetting in Galicia and present the results of classifying these causes and motivations for homogenous geographical areas. The final section recapitulates the main points, and offers some conclusions.

2 Spatial clustering of forest fire causes and motivations

2.1 Data and methodology

This analysis uses wildfire data from Galicia, 1978-2006, provided by the General Statistics of Forest Fires compiled by the Spanish Forest Service and Consellería do Medio Rural (Xunta de Galicia). Spain has had a standardized forest fire database since 1968, making it one of the world's most comprehensive wildfire datasets. Although the reports have undergone several changes during all this years, the uniformity and continuity of the information has always been maintained (Martín [2]). The current version of these reports gathers general information regarding the fire (e.g. area burned, date and time of ignitiation, climatic conditions, causes and motivations, fire fighting measures) and detailed information on the forestland affected (e.g. ownership, forest biomass of the area, and estimated losses). Detailed information on the causes and motivations of intentional fires were not fully captured in these reports until the end of the 1980's. Therefore, the data used in this study range from 1988 to 2006 (19 years). This time period includes problematic events such as those experienced in 1989, when Galicia registered more than 8,000 wildfires, affecting about 200,000 ha.

Fires from the database were assembled into a dataset of counts of wildfires at the level of the municipality. For each municipality, counts of wildfires by causes and motivations were assembled. Fire policies and prevention efforts are organized at the level of forest districts. There are 19 forest districts, each of which compromises several municipalities. There are therefore too heterogeneous. We opt for the municipalities as the unit of our analysis. (e.g. Martínez and Chuvieco [14]). There are currently 315 municipalities in Galicia, with a mean size of 9,300 ha. Our number of observations however is 313, because several municipalities were studied together given that these were seggregated during the study period. This is the case of Burela and Cervo, seggregated in 1995, and Illa de Arousa and Vilanova de Arousa, seggregated in 1997. In order to be able to compare between information of different municipalities, a normalisation was done, expressing fire counts relative to area.

The data were analyzed using cluster analysis (Romesburg [18]). This allows us to classified the Galician municipalities in different groups in order to find data associations and relationships in such a way that the degree of similarity (or homogeneity) among municipalities within a group is stronger than among municipalities of different groups based on the causes and motivation variables. In general, the greater the homogeneity within a group and the greater the difference between groups, the more distinct the clustering. The most common technique used is hierarchical clustering. However, this technique is unsuitable for this study because it has an elevated number of observations. We use *k*-means clustering, which involves splitting the municipalities in a predetermined *k*-number of groups. We compute a stopping-rule index, in particular the Calinski-Harabasz (Calinski and Harabasz [19]) pseudo-F index, for each cluster solution to determine the *k*-number of cluster. Higher values of this index indicate more distinct clusters. Discriminant analysis was also applied to check and improve the classification obtained with the application of the cluster technique. Discriminant analysis classifies municipalities into one of several mutually exclusive groups, which are given by the results previously obtained with cluster analysis, based on their values for the a set of predictor variables (i.e., the causality variables). Statistical software used was STATA 11® and SPSS V. 15.

Municipalities are classified according to the forest fire causes and underlying motivations identified in the Galician forest fire reports. These reports distinguished among six general causes of forest fires, which could be dissagregated into about forty different specific causes. In addition, human-intentional wildfire ignitions are classified according to twenty-four social motivations. Table 1 presents the classification of the causes following the forest fire reports. It only includes the relevant causes for the Galician case study (i.e., those causes for which there are observations available for Galicia in the study period). Agricultural and rangeland burning are either negligent or intentional forest fires depending on weather or not the responsible person undertakes the neccesary cautious measures established by law and has the corresponding burning permission. This means that firesetting in agricultural and livestock activities is not always considered as a criminal use of fire. In contrast, the rest of motivations of intentional fires in Table 1 are illegal activities because they cannot be authorized in any case.

2.2 Results

In the study period (1988-2006), 82% of the fire ignitions are classified as intentional while the fires caused by negligence and accidents only represent approximately 5%. Fires with unknown causes is 9%. These fires have been decreasing significantly since the begining of 1990's in Spain (Castedo et al. [20]), presumably as a result of efforts to overcome this shortage. However, the proportion of the unknown caused-fires in Galicia is, in fact, among the lowest of Spanish regions, where this percentage can be over 40% (Martín [2]). In the negligence and accidental forest fires, agricultural burning and forestry activities such as elimination of shrubs for afforestation are the main causes of negligent fires, 14% and 12% respectively. Rangeland burnings to enable pasture regrowth and rubbish tip escapes are next in importance, at 7% each. Less common

Table 1: Causes classification based on Galician forest fire reports.

Forest fire causes	Specific type of cause	
Lightning		
Negligence and accidental I	Agricultural burning	(a)
	Rangeland burning	(b)
	Forestry work	(c)
	Burning bush shrub	(d)
	Bonfires	(e)
	Smokers	(f)
	Debris fire escapes	(g)
	Other negligent causes (e.g. fireworks)	(h)
Negligence and accidental II	Railway	
	Electric lines	
	Motors and machines (e.g. accidents, heavy and light vehicles)	
	Military manuevers	
Intentional	Type of Motivations: Agricultural burning; rangeland burning; pyromania; hunting; vandalism; get salaries; change land use; revenge; dispute against a fines; resentment against reforestation; drive away animals; watch forest fire fighting; distract the police; rituals; cancel contracts with administration; resentment against subsidies; etc.	
Unknown		
Reproduced		

remaining causes include campfire escapes, smoking, electric power lines, and railway lines. The so-called other negligent causes not clearly identified account for 33% of all intentional wildfires

Intentional fires are a serious concern for policy makers, because they cause great damage. They often occur in inaccesible locations, and several of them can start at the same time in different places. For the study period, the percentage of intentional fires with a registered motivation is 54%, while the rest has unspecified motivations (13%). The motivations of people that start these types of fires vary widely. The most important include agricultural and shrubland burnings (31%). Intentional rangeland burning to regrow pastures is the second most commonly identified motivation (11%), while pyromania is the third (9%). The rest of the cases (4%) are divided into motivations related to hunting, revenge due to land use and property, vandalism, rejection of land use limitations derived from afforestation and protected areas, attempts to modify land use into building land, low timber prices, etc.

In summary, the variables selected to classified Galician municipalities according to the forest fire causes and intentional forest fires motivations are the following:

- Percentage of fires caused by negligences and agricultural activities: (a)+(b)+(c)+(d)
- Percentage of fires caused by other negligences and accidental causes: (e)+(f)+(g)+(h)
- Percentage of intentional forest fires motivated by agricultural burning.
- Percentage of intentional forest fires motivated by rangeland burning.
- Percentage of intentional forest fires motivated by pyromaniacs.

Other potential variables of interest that could also help to define the causes and motivations of forest fires were excluded from the statistical analysis either for (i) representing an insignificant proportion of the total fires occurrances (e.g., percentage of fires caused by negligence and accidental causes II); or (ii) for having a high correlation ($r \geq \pm 0.5$) with some of the variables considered above (e.g., the percentage of unknown causes is highly correlated with intentional forest fire).

The cluster analysis shows that Galician municipalities may be classified according to the causes of forest fires in six distinct clusters. After carrying out the discrimant analysis using the same causality variables, 92.4% of the municipalities are correctly classified in the original cluster analysis. This implies that only 25 municipalities had their classifications modified through discrimant analysis. There is therefore a high goodness-of-fit in the discriminant classification function. The final number of municipalities in each cluster is presented in Table 2. This table also shows the proportion of municipalities in each cluster with respect to the total number of analyzed spatial units. Other features displayed include the proportion of the forest area, burned area, and number of forest fires in each cluster with respect to Galician total forest area, total burned area and total number of forest fires, respectively, in the nineteen years of our study period.

Table 2: Number of municipalities in each cluster.

Cluster	Municipalities	% total municipalities	% Galicia forest area	% Galicia burned area	% Galicia nº of forest fires.
1	36	11.5	6.35	11.26	17.64
2	37	11.8	13.76	16.43	16.69
3	92	29.4	28.27	31.11	29.48
4	110	35.1	36.13	26.65	24.83
5	17	5.4	9.07	9.33	9.85
6	21	6.7	6.41	2.23	1.52
Total	313	100	100	100	100

Table 3: Mean and standard deviation of general properties of forest fires for the different clusters identified*.

Variables	Clusters					
	1	2	3	4	5	6
Forest area (ha)	3,594 (2,317)	7,577 (3,756)	6,261 (4,815)	6,630 (5,833)	10,871 (5,185)	6,221 (4,258)
Burned area (ha)	2,502 (2,161)	3,551 (2,524)	2,704 (2,569)	2,136 (1,990)	4,387 (2,802)	849 (877)
Nº of forest fires	870.8 (697.7)	801.8 (534.4)	569.6 (394.1)	397.6 (273.5)	1029.5 (632.0)	129 (99)
Risk index ***	0.27 (0.16)	0.11 (0.05)	0.11 (0.07)	0.08 (0.06)	0.10 (0.06)	0.02 (0.01)
Seriousness index ***	65.8 (43.2)	44.3 (22.8)	45.4 (31.8)	37.3 (23.6)	41.3 (23.9)	13.6 (9.4)
% Forest fires due to natural causes	0.20 (0.29)	0.80 (1.0)	0.8 (1.2)	2.2 (2.5)	0.8 (0.7)	1.8 (3.4)
% Forest fires due to agricultural negligence	2.3 (1.3)	1.3 (1.4)	3.6 (2.7)	2.8 (2.5)	2.8 (2.5)	16.7 (8.1)
% Forest fires due to other negligence or accidents	2.8 (1.8)	1.5 (1.0)	3.1 (2.3)	2.8 (2.2)	2.2 (2.2)	6.0 (3.5)
% Intentional forest fires	74.1 (8.9)	88.1 (4.8)	76.4 (10.3)	70.5 (13.1)	86.0 (8.3)	45.8 (15.5)
% Intentional forest fires motivated by agriculture burning**	13.1 (3.4)	37.6 (11.4)	13.0 (4.4)	2.4 (1.8)	16.1 (8.4)	3.6 (3.3)
% Intentional forest fires motivated by rangeland burning **	1.9 (2.8)	11.5 (4.6)	4.8 (4.6)	1.2 (1.5)	31.2 (10.5)	5.0 (5.8)
% Intentional forest fires motivated by pyromaniac behavior **	17.9 (6.4)	0.6 (0.6)	1.9 (2.6)	0.4 (0.8)	0.4 (0.8)	0.4 (1.1)
% Forest fires due to unknown causes	16.9 (7.5)	7.2 (3.8)	14.2 (7.6)	19.9 (11.9)	6.4 (4.0)	26.8 (14.7)

* Standard deviation in brackets. ** Calculate based on the total number of intentional forest fires. *** The risk index captures the density of forest fire per ha and seriousness index shows the percentage of burned area with respect to the total forest area.

Table 3 presents the mean and standard deviation of the forest fire causes and motivations for each cluster. It includes other relevant information to characterize the resulting clusters. For example, forest area, burned area, number of forest fires, risk index, and seriousness index.

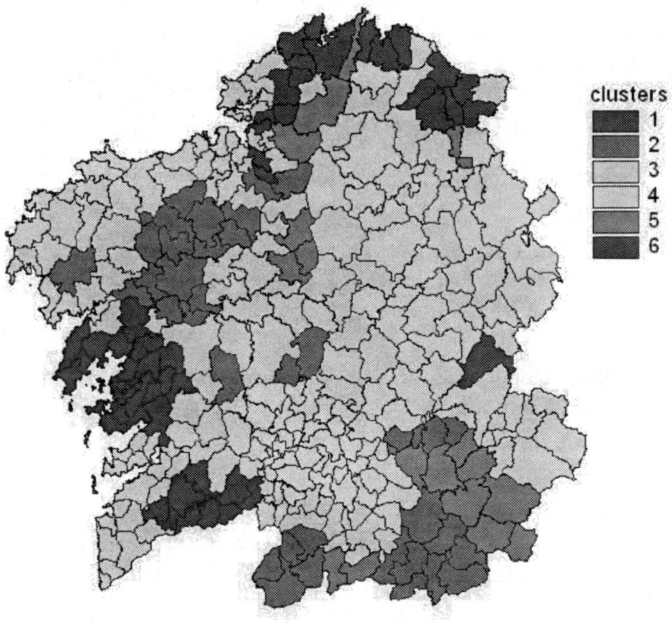

Figure 1: Spatial distribution of homogenous municipalities attending to causality properties of forest fires in Galicia (1988-2006).

Clusters present groups of municipalities with homogenous characteristics in relation to the causes and underlying motivation of forest fires. We expect the resulting classification to have not only a coherent typology in terms of the causality variables, but also a geographically meaningful dispersion in the Galicia territory. In other words, the groups of homogeneous municipalities are expected to have some cluster behavior in a spatial dimension. Figure 1 shows that there is a high level of spatial clustering in the homogeneous municipality groups derived from this analysis.

Based on these results, it is possible to classify Galician municipalities in four different types according to the causes and motivations of forest fires ignitions. These include:

Type 1 (clusters 2 and 5): Dominated by intentional forest fires associated with agricultural and rangeland activities.

In these Galician municipalities, there is a high percentage of intentional forest fires, nearly 90% of all reported fires, and the within-group dispersion of this variable is very small (Table 3). Forest fires that have unknown causes are the lowest of the whole region (6%-7%) (Table 3). Spatially, this type of municipality is clustered in two main blocks: one in the southeast, and a more dispersed one in northern Galicia. Intentional agricultural burning and rangeland burning dominate cluster 2 and 5, respectively. Uncontrolled burning of shrub and residues of agricultural activities motivate 37% of the intentional forest fires in municipalities of cluster 2 but just 16% of those fires in municipalities of cluster 5. On the other hand, rangeland burning to increase pasture productivity explains 11% of intentional fires in cluster 2 but 31% of those fires in municipalities of cluster 5 (Table 3). Intentional agricultural burning fires in cluster 2 represent 40% of the total Galician intentional forest fires; and intentional rangeland burning fires in cluster 5 represent 43% of the total Galician intentional forest fires. Prevention policies in municipalities in *Type 1* should mainly focus on the control of firesetting related to illegal agricultural and livestock burning.

Type 2 (cluster 1): Dominated by intentional forest fires caused by pyromaniacal behavior

Municipalities with this typology are located mainly in the South coast of Galicia. Here, intentional forest fires are highly relevant, with a 74% of the total fires within this group of municipalities and a 17% of total fires in Galicia (Table 3). Intentional forest fires caused by agricultural and livestock motivations are underrepresented. In this zone, only 13% and 2% of the intentional forest fires are associated with agricultural and livestock activities, respectively. Pyromaniacs explain 18% of the intentional fires in these municipalities, which represent 77% of the total intentional fires in Galicia by this motivation. Therefore, this area concentrates the malicious acts of pyromaniacs, because more than three-quarters of this type of event in Galicia occur here. Note that forest area in this zone represents 6% of the Galician forest area, but 11% of the burned area, and 17% of the total number of fires (Table 2). Risk and seriousness indices are the highest in the region (Table 3).

Type 3 (clusters 3 and 4): Dominated by intentional forest fires with unspecified motivations

It includes clusters 3 and 4, which have the highest number of municipalities (Table 2). They represent 64% of municipalities. In the municipalities of *Type 3*, the percentage of intentional forest fires is relatively high, with a mean value of 70%. However, there is higher within-group dispersion than in other groups. In cluster 3, intentional forest fires with agricultural, livestock and pyromaniacal motivations represent 13%, 5% and 2%, respectively. In cluster 4, these percentages take smaller values with 2.5%, 1.2% and 0.4%, respectively (Table 3). This indicates that only a small percentage of intentional forest fires in this zone can be explained with the motivations considered in this cluster analysis. We can therefore conclude that motivations that explained this type of forest fires are other or unspecified motivations. The properties of this zone indicate that in an important part of the Galician territory (cluster 3 and mainly cluster 4), stepped up efforts to identify motivations underlying intentional forest fires are needed as prerequisites to the development effective prevention policies.

Type 4 (cluster 6): Low weight of intentional forest fires, and higher importance of negligence and unknown causes

In the municipalities of this typology, the mean number of forest fires and the mean percentage of intentional forest fires are the lowest of Galicia (46%). Only 8% of these intentional fires are apparently motivated by agricultural and livestock activities (Table 3). However, the percentage of forest fires caused by negligence and accidental is on average 23%. These fires are caused mainly by careless behavior in agricultural activities (17%). There is also a high percentage of fires with unknown causes (27%) relative to the total number of fires ignited in these municipalities (Table 3).

3 Discussion and conclusions

This research has analyzed forest fires in Galicia in the period 1988-2006 using information from individual fires reports. Based on multivariate techniques, Galician municipalies are classified in six clusters. This allows us to divide the municipalities of this region into four distinct types according to their prevalances of the various causes and underlying motivations in forest fire occurence. *Type 1* is dominated by intentional fires motivated by malicious or illegal agricultural and rangeland burning. *Type 2* is characterized by pyromaniacal firesetting and a high fire frequency and burned area per forest area (i.e., it has the highest risk and seriousness indices). Intentional fires with unspecified motivations dominate *Type 3*. Finally, *Type 4* is characterized by fire ignitions associated with careless behavior and accidents, and with unknown general causes. The proportion of intentional fires in *Type 4* is the lowest, compared to the rest of Galicia. A straight forward extension of this work is to analyze this typology and zonification of forest fires causes and underlying motivations at a finer spatial scale (e.g. parish level with 3,801 parishes with an average size of 780 ha). Preliminary work shows no sensitivity of the results presented here to the spatial scale.

The analysis of the causes of fire occurrence is not an easy task because of the climatic, ecological, and socioeconomic variables that may simultaneously affect the probability of these events. For example, intentional fires often occur in very dry periods and in locations with high biomass inflamability. More detailed future analyses, which account for the interactions among natural and social factors and seasonality, and that explore how clusters differ when data are refined to smaller spatial units, could advance our understanding of such processes. Nevertheless, our study of the different causality typologies for the Galician municipalities based on cluster analysis shows that there are distintive spatial patterns of behaviors in forest fires ignitiations. This information could help in the development of more spatially targeted efforts of fire prevention and campaigns to enhance the awareness of local populations about the risks associated with the use and abuses of fire and the potential legal consequences of such activities.

Acknowledgements

This work was funded by Conselleria de Economía e Industria-Xunta de Galicia (Project 09SEC011201PR). The authors would like to thank the Spanish Forest Service and Consellería do Medio Rural (Xunta de Galicia), which provided the forest fire database. An earlier version of this paper was presented at 5° Congreso

Forestal Español, held in Ávila (Spain), 23-25 September 2009. We would like to thank participants at this conference for their comments and suggestions.

References

[1] European Commission. *Forest fires in Europe 2008*. EUR-23971EN. Joint Research Centre. Institute for Environment and Sustainability. Report nº 9. 88 p., 2009.

[2] Martín, P. (Coord.); *Towards methods for investigating on wildland fire causes.* EUFIRELAB. EVR1-CT-2002-40028. Deliverable D-05-02. 54 p, 2004.

[3] Gamst, F., *Peasants in Complex Society*. New York: Holt, Rinehart, and Winston, 1974.

[4] Doolittle, M.L., & Lightsey, M.L., Southern woods-burners: A descriptive analysis. Research Paper SO-151, USDA Forest Service. 6 pp., 1979.

[5] Vélez, R., Incendios forestales y su relación con el medio rural. *Revista de Estudios Agrosociales*, 136, pp. 195-224, 1986.

[6] Kuhlken, R., Settin' the woods on fire: Rural incendiarism as protest. *The Geographical Review*, 89, pp. 343-363, 1999.

[7] Prestemon, J.P. & Butry, D.T., Wildland arson management. *The Economics of Forest Disturbances: Wildfires, Storms, and Invasive Species,* ed. T.P. Holmes, J.P. Prestemon and K.L. Abt, Springer. Dordrecht, The Netherlands, pp. 123-147., 2008

[8] Bertrand, A.L., Hefferman, W.D., Welch, G.D. & O'Carroll, J.P., *Attitudinal patterns in a forest area with high incendiarism.* Baton Rouge. Louisiana Agricultural Experiment Station Bulletin No. 648, 1970.

[9] Bertrand, A.L. & Baird, A.W., *Incendiarism in Southern Forests: A Decade of Sociological Research.* Bulletin 638. Mississippi State University and Southern Forest Experiment Station, U.S. Forest Service, 1975.

[10] Maingi, J.K., & Henry, M.C., Factors influencing wildfire occurrence and distribution in eastern Kentucky. *International Journal of Wildland Fire* 16, pp. 23-33, 2007.

[11] Prestemon, J. P. & Butry, D.T., Time to burn: Modeling wildland arson as an autoregressive crime function. *American Journal of Agricultural Economics,* 87(3), pp. 756-770, 2005.

[12] Butry, D.T. & Prestemon, J.P., Spatio-temporal wildland arson crime functions. Paper presented at the Annual Meeting of the American Agricultural Economics Association, July 26-29, 2005, Providence, Rhode Island. 18 p. Available at: http://agecon.lib.umn.edu/cgi-bin/pdf_view.pl?paperid=16442&ftype=.pdf

[13] Genton, M.G., Butry, D.T., Gumpertz, M. & Prestemon, J.P., Spatio-temporal analysis of wildfire ignitions in the St. Johns River Water Management District. *International Journal of Wildland Fire* 15, pp. 87-97, 2006.

[14] Martínez, J. & Chuvieco, E., Tipologías de incidencia y causalidad de incendios forestales basadas en análisis multivariante. *Ecología* 17, pp. 47-63, 2003.

[15] Vázquez, A. & Moreno, J.M., Patterns of Lightning and People-Caused Fires in Peninsular Spain. *International Journal of Wildland Fire*, 8(2), pp. 103-115, 1998.

[16] Amatulli, G., Peréz-Cabello, F., & de la Riva, J., Mapping lightning/human-caused wildfires occurrence under ignition point location uncertainty. *Ecological Modelling,* 200, pp. 321-333, 2007.

[17] Benavent-Corai, J., Rojo, C., Suárez-Torres, J. & Velasco-García, L., Scaling properties in forest fire sequences: The human role in the order of nature. *Ecological Modelling*, 205, pp. 336-342, 2007.

[18] Romesburg, H.C., *Cluster analysis for researchers*. Lulu Press North Carolina, 2004.

[19] Calinski, T. & Harabasz, J., A Dendrite method for cluster analysis. *Communications in Statistics,* 3(1), pp. 1–27, 1974.

[20] Castedo Dorado, F., Juarez Relaño, I., Ramírez Cisneros, J., Ruiz Pérez, I.; Rodríguez Rodríguez, C. & Velez Fraile, L., Utilidad del análisis de la estadística de incendios en las estrategias de prevención y extinción. Un caso de estudio. *Proceedings of the 4th International Wildland Fire Conference*, Seville, Spain, May 13-17, 2007.

Evaluation of the FCCS crown fire potential equations in Aleppo pine (*Pinus halepensis* Mill.) stands in Greece

M. D. Schreuder[1], M. D. Schaaf[1] & Da. V. Sandberg[2]
[1]*Air Sciences Inc., Portland, Oregon USA*
[2]*USDA Forest Service (Emeritus), Corvallis, Oregon, USA*

Abstract

This paper evaluates the conceptual crown fire potential equations developed by Schaaf *et al* against observations and modelling results in Aleppo pine (*Pinus halepensis* Mill.) stands in Greece. The equations, integrated into the Fuel Characteristic Classification System (FCCS) in the United States, are currently used to rank the potential for passive or active crowning across a diverse set of wildland fuelbeds. The framework is based on an extension of the work by Van Wagner and Rothermel but introduces several new physical concepts to the modelling of crown fire behaviour, including the reformulated Rothermel surface fire modelling concepts proposed by Sandberg *et al.* A sensitivity analysis comparing the FCCS Torching Potential (*TP*) and Active Crown Fire Potential (*AP*) against field observations and CFIS modelling outputs has produced encouraging results, suggesting that the FCCS crown fire potentials might be a useful tool for fire managers in the Mediterranean region to consider when evaluating the relative behaviour of crown fires in vegetated canopies.
Keywords: crown fire, crown fire potential, modelling, fire behaviour, canopy, torching, active crown fire, FCCS, aleppo pine, Mediterranean.

1 Introduction

The Fuel Characteristic Classification System (FCCS), [6], is a system that describes the physical characteristics of any North American wildland fuelbed, regardless of complexity, and facilitates comparisons of the physical fuelbed characteristics and fire potentials. FCCS enables the user to assess the absolute and relative effects of fuelbed differences due to natural events, fuel management

WIT Transactions on Ecology and the Environment, Vol 137, © 2010 WIT Press
www.witpress.com, ISSN 1743-3541 (on-line)
doi:10.2495/FIVA100231

practices, and successional changes over time. The differences can be expressed as native physical differences, such as changes in loadings and arrangements of fuelbed components; or as changes in the potential fire-related effects, such as fire behaviour or fuel consumption [5]. Until recently in FCCS, the lack of a broadly applicable and physics-based crown fire model capable of utilizing the comprehensive description of fuelbeds in FCCS hampered comparisons of the crown fire potentials among environmentally diverse fuelbeds. FCCS does not require specific predictions of crown fire behaviour across the range of fire environments but does require a relative ranking of crown fire potential over the range of wildland fuelbed characteristics. To this end, a conceptual model of crown fire initiation and spread was developed and eventually integrated into the system [1].

The body of literature advancing the science of crown fires, [7–17], shows that the potential for crown fire initiation and spread does not depend on any single element of the fuel complex, fire weather environment, or topography, but rather from combinations of interrelated variables, including: surface fire intensity, canopy closure, crown density, the presence of ladder fuels, height to base of the combustible crown, crown foliar moisture content, and wind speed. The FCCS crown fire potentials are based on an updated semi-empirical model that describes crown fire initiation and propagation in vegetative canopies based on the work by Van Wagner [2] and Rothermel [3], but updated with additional physical concepts for modelling crown fire behaviour derived from the reformulated Rothermel [4] surface fire equations proposed by Sandberg *et al* [5]. This crown fire modelling framework is conceptual in nature. It has had limited testing against independent data sets [1], and its use is currently limited to assessing the crown fire potential of FCCS fuelbeds. Additional refinement and verification are needed before the FCCS crown fire model can be considered for wider application. In an earlier effort the FCCS crown fire equations were evaluated against crown fire observations in black spruce (*Picea mariana* (P. Mill.) B.S.P.) [1]. This paper provides additional evaluation against pine stands characteristic for the eastern Mediterranean. Specifically, a sensitivity analysis was performed for Aleppo pine (*Pinus halepensis* Mill.) stands in Greece [18, 19], and crown fire behaviour was compared with reported modelling results.

2 FCCS crown fire potentials

The FCCS crown fire modelling framework ranks the relative potential for crown fire initiation and spread in natural fuelbeds based on a set of actual and inferred characteristics. It draws upon published models and results from crown fire experiments by others, personal observations of crown fires, and conversations with fire managers. This model is intended to objectively assess, on a relative scale, the probability of experiencing torching or active crown fire spread in any FCCS fuelbed. Currently applied crown fire models, [2, 7, 13–17], are largely empirically based and appropriate only when applied to the range of stand structures and fire behaviours observed. While these models are very useful under certain circumstances, they do not provide the broad conceptual framework or applicability necessary to compare the crown fire potential within families of dissimilar fuelbeds. For that, the authors sought a more universal approach through the development of this conceptual model.

2.1 General form

The general form of the FCCS crown fire potential (CFP) equation is:

$$CFP = \max\left(TP, AP\right) \tag{1}$$

where *TP* is the torching potential and *AP* is the active crown fire potential. Both are dimensionless, ranging in value from zero to 10. *TP* is the potential for a surface fire to spread into the canopy as single tree or multiple tree torching. If *TP* is greater than one, then torching is possible. *TP* is defined as the scaled crown fire initiation term, I_C (dimensionless, range zero to 10):

$$TP = c_{TP} \cdot I_C \tag{2}$$

Here, c_{TP} is a scaling function limiting *TP* to within the range of zero to 10. Fuelbeds with I_C values greater than ten are assigned a *TP* of ten. Fuelbeds with I_C values less than ten are scaled from zero to ten.

AP is the potential for a surface fire to spread into the canopy as an active crown fire. If *AP* is greater than zero, than active crown fire spread is possible. *AP* is computed as the scaled product of four terms:

$$AP = c_{AP} \cdot I_C \cdot F_C \cdot R_C \tag{3}$$

where c_{AP} is a scaling function that limits *AP* to a range of zero to 10 (dimensionless), I_C is the crown fire initiation term, F_C is the crown-to-crown flame transmission term (dimensionless, range zero to one), and R_C is the crown fire spread-rate term (m/min, range one to >100 m/min).

Fuelbeds with the product of $I_C \cdot F_C \cdot R_C$ greater than 10 are assigned an *AP* of 10. Fuelbeds with the product of $I_C \cdot F_C \cdot R_C$ less than 10 are assigned an *AP* scaled from zero to 10.

The development of the I_C, F_C, and R_C terms are described in Schaaf *et al* [1] and outlined below.

2.2 Crown fire initiation term

Following Van Wagner [2], crown fire initiation is expected when the surface fireline intensity, I_B (kW/m), exceeds the critical fireline intensity I' (kW/m); that is, when $I_B/I' > 1$. This ratio yields:

$$\frac{I_B}{I'} = \frac{16.667\, I_R\, t_R\, R}{\left[CBH\left(460 + 25.9\, FMC\right)\right]^{3/2}} \tag{4}$$

where I_R is reaction intensity (kJ/m^2-min), t_R is residence time in minutes, R is the forward rate of spread of the fire (m/min), *CBH* is canopy base height (m), *FMC* is foliar moisture content of canopy fuels (%), and 16.667 is a constant of proportionality that produces the correct unit conversion (minutes to seconds, among others).

Eqn. (4), expressed in terms of FCCS variables with the residence time set to the inverse of the surface potential reaction velocity, defines the FCCS crown fire initiation term, I_C:

$$I_C = \frac{16.667\, I_{R.FCCS} \left(1/\Gamma'\right) R_{FCCS.S}}{\left[\left(gap/ladder\right)\left(460 + 25.9\, FMC\right)\right]^{3/2}} \tag{5}$$

where Γ' is the potential reaction velocity (1/min) of surface fuels from Rothermel [4], $R_{FCCS.S}$ is the surface fire spread rate from FCCS (m/min), *gap* is the physical separation distance between the top of the surface fuel layer and the bottom of the combustible canopy layer (m), and *ladder* is an heuristically assigned value representing the presence and type of ladder fuels sufficient to act as a vertical carrier of fire to the canopy base (default *ladder* = 1, meaning no ladder fuels are present). These terms are all defined in [5]. While the traditional Van Wagner [2] equation bases the calculation of I' on CBH, FCCS expresses I' in terms of the vertical gap between the top of the surface fuelbed layer and the bottom of the combustible canopy layer, with adjustments related to the abundance of combustible ladder fuels. The validity of this modification will be evaluated in future model validation efforts.

I_C is evaluated along a continuum ranging from zero to infinity. The higher the I_C value, the greater is the potential for initiating a crown fire. This is the same equation set used in [7] except that they structured the equations in a manner that established mid flame wind speed as the principal variable, whereas we have structured the equations to evaluate the initiation potential across a range of fuelbeds with different surface reaction intensities, and rates of spread at a variable benchmark wind speed (default mid flame wind speed is 107 m/min, or ~6.4 km/hr).

2.3 Crown-to-crown flame transmission term

The FCCS crown-to-crown flame transmission term (F_C) is a dimensionless measure of the capacity to transfer flames through the canopy based on leaf area index (*LAI*), wind speed, and horizontal continuity of tree crowns. For canopies above some threshold *LAI*, the greater the wind speed the greater the *effective* horizontal continuity of tree crowns and the greater the crown-to-crown transmission of flames. And the higher the transmission rate, the greater is the potential to sustain an active crown fire. Torching is not affected by the horizontal continuity of tree crowns.

This new conceptual term is proposed as a replacement for the model originally proposed by Van Wagner [2], which determines whether an active crown fire will occur by comparing the estimated crown fire spread rate with a critical spread rate required to sustain an active crown fire. Although practical and widely used, Van Wagner's model [2] assumes that the canopy is horizontally uniform and continuous. It does not explicitly account for spacing between tree crowns nor does it consider the effect of increasing mid-canopy wind speed in reducing the effective spacing. Moreover, application of the Van Wagner model relies on an estimate of crown-fire spread rate based on a limited correlation developed by Rothermel [3]. The new approach in FCCS is less supported by observations than that developed by Van Wagner [2] and Rothermel [3] but is more physically intuitive. Additional testing of this modelling concept is needed.

The FCCS crown-to-crown flame transmission term, F_C, is defined as follows:

$$F_C = \begin{cases} 0, & \text{for } LAI < TLAI \\ \dfrac{\max\left(0, [COV \times WAF] - TCOV\right)^{0.3}}{\left([100 \times WAF] - TCOV\right)^{0.3}}, & \text{for } LAI \geq TLAI \end{cases} \tag{6}$$

where LAI is leaf area index (m^2/m^2), $TLAI$ is threshold LAI for active crowning (m^2/m^2), COV is total percentage cover of tree crowns (i.e., percentage canopy cover, or percentage ground area covered by tree crowns) (dimensionless), WAF is a canopy wind speed adjustment factor (dimensionless), and $TCOV$ is threshold percentage canopy cover (dimensionless) required to propagate an active crown fire when $WAF = 1$ ($TCOV = 40$). $TLAI$ was estimated based on Van Wagner's [2] empirical relationship that describes the interaction between canopy bulk density and the minimum spread rate needed to sustain an active crown fire. The resulting formulation is:

$$TLAI = \frac{CBD_{critical}\ \sigma\ D_C}{\rho_p} \tag{7}$$

where $CBD_{critical}$ is the canopy bulk density (kg/m^3) required to sustain an active crown fire, σ (m^2/m^3) is the surface-area-to-volume ratio of foliage elements, D_C is the mean canopy depth (m), and ρ_p is the particle density (kg/m^3). The development of these terms is described in [1].

Eqn. (6) assumes that a relatively continuous canopy is required for efficient crown-to-crown heat transfer. This validity of this assumption should be evaluated in future field studies.

2.4 Crown fire spread-rate term

The FCCS crown fire spread-rate term, R_C, is a new, physically-based mathematical approach for estimating the crown fire spread rate using the reformulated Rothermel surface fire spread rate [5], adapted to vegetative canopies. The reformulated Rothermel spread rate is the ratio of a surface-fuel heat source term acting to accelerate the fire spread (numerator), and a heat sink term representing the sum of individual component heat sinks acting to retard the fire spread (denominator). The heat source term includes formulations for reaction intensity, propagating flux ratio, and an acceleration factor for wind. The heat sink terms include various physical fuelbed characteristics, including fuel area index, ignition thickness, heat of pre-ignition, and fuelbed thickness, among others. These terms are described in detail in Sandberg *et al* [5].

For active crown fires, the combined reaction intensity produced by the flaming combustion at the surface as well as from the flaming canopy fuels drives the forward heating of the fuels and associated fire spread rate. In the FCCS crown fire potential framework, the reformulated Rothermel spread rate has been applied to canopies in a manner similar to its application to the surface fuelbed, with corresponding terms for both the fuel heat source and sink terms based on the unique characteristics of each FCCS fuelbed including a vegetative canopy.

Because of its complexity, a comprehensive description of the crown fire spread-rate term is beyond the scope of this paper. A detailed description is in [1].

3 Sensitivity analysis

3.1 Methods

The FCCS crown fire equations were tested using data and observations from Aleppo pine (*Pinus halepensis* Mill.) stands in Greece [18, 19]. In a first comparison, five FCCS pine fuelbeds [20] were selected and evaluated for their potential to represent even- and uneven-aged Aleppo pine stands. The two most representative FCCS fuelbeds were: Fuelbed164 (sand pine forest) and Fuelbed 282 (loblolly pine forest) for even- and uneven-aged stands, respectively (Table 1 [21]).

The FCCS models were selected based on the correspondence between critical parameters in crown fire modelling; specifically, crown fuel loading (*CFL*), crown bulk density (*CBD*), crown base height (*CBH*), and surface fuel loading (*SFL*). Although the *CBH* of the FCCS stands is somewhat lower than those reported in Table III of the original paper [19], they are comparable to the *CBH* in typical Aleppo pine stands reported in Table I [19]; specifically, 3.1 m. The two FCCS models were also run against the CFIM/CFIS equations, yielding results that are roughly equivalent to those reported (Table V [19]).

Table 1: Summary of stand and fuel characteristics for Aleppo pine stands and FCCS fuelbeds.

	Even-aged Aleppo pine	FCCS #164	Uneven-aged Aleppo pine	FCCS #282
Canopy closure (%)	76[19]	75	69[19]	85
Stand density (n/ha)	700[19]	1482	697[19]	988
Stand height (m)	15.7[19]	8.5	20.9[19]	19.8
Crown fuel loading (kg/m^2)	1.4[19]	1.1	1.4[19]	2.3
Crown bulk density (kg/m^3)	0.18[19]	0.18	0.16[19]	0.14
Crown base height (m)	5.3[19]	2.7	3.9[19]	3.0
Surface fuel load (t/ha)	2.35[18]	1.95	2.35[18]	3.76

In the next step, the two FCCS models were tested for their sensitivity to several key input parameters. Specific input parameters tested were the effects of fire weather class, fuel strata gap (*FSG*), and flammability of the fuels. Wind speed conversions between the 10-meter (U_{10}, km/hr) and mid-flame (1-m wind speed, U_1) wind speeds were assumed a logarithmic vertical wind profile, resulting in U_{10}~$2*U_1$ (similar to the conversion in [1]). The fire weather classes were approximated by varying the U_{10} and estimated fine fuel moisture content (EFFM, %). Four fire weather classes were tested: low (U_{10}=10, *EFFM*=12.5), moderate (U_{10}=20, *EFFM*=9.375), high (U_{10}=30, *EFFM*=6.25), and extreme (U_{10}=30, *EFFM*=3.125). These levels are similar to those reported in CFIM/CFIS model runs for Aleppo pine stands [19]. The effect of *FSG* (as defined by FCCS) was tested by calculating the physical distance between the top of the stand understory (Table V, [19]), and the *CBH* (Table III, [19]). The resulting *FSG* values were then run in FCCS and the effects on potential (crown) fire behaviour evaluated. The specific output variables consisted of *TP* (eqn. 1), *AP* (eqn. 2), and the FCCS-based surface and crown fire rates of spread, $R_{FCCS.S}$ and $R_{FCCS.C}$, respectively.

3.2 Results

Table 2 summarizes the sensitivity of crown fire perimeters to different fire weather classes, assuming a *FSG* of 0.9. This *FSG* can either be achieved through the physical gap between the *CBH* and the top of the understory, or, alternatively, through the presence of sufficient ladder fuels to bridge the gap between the two. For both fuelbeds, the *AP* values are greater than zero, indicating a potential for active crown fire (Table 2). However, in all but the extreme weather conditions the fire stays at the surface (*TP*<1; Table 2). Only in Fuelbed 164, with a slightly lower *CBH*, does the fire crown under the extreme fire weather scenario. Fuelbed 282 tends to have somewhat higher fire rates of spread, both in terms of $R_{FCCS.S}$ and $R_{FCCS.C}$ (Table 2), primarily due to higher surface loading and crown fuel loading (Table 1). For both fuelbeds, *TP* approximately doubles between low and extreme fire weather conditions, whereas *AP* and $R_{FCCS.C}$ approximately triple in value (Table 2).

Table 2: FCCS-predicted fire behaviour by fire weather class (*FSG*=0.9 m).

	Low	Moderate	High	Extreme
	FCCS Fuelbed 164			
TP	0.7	0.8	0.9	1.3
AP	3.3	4.3	7.3	10.0
Fire type	Surface	Surface	Surface	Active crown
$R_{FCCS.S}$ (m/min)	1.8	1.9	2.1	2.5
$R_{FCCS.C}$ (m/min)	1.8	1.9	2.7	3.1
	FCCS Fuelbed 282			
TP	0.5	0.5	0.6	0.9
AP	3.4	7.8	10	10
Fire type	Surface	Surface	Surface	Surface
$R_{FCCS.S}$ (m/min)	2.2	2.3	2.5	3.1
$R_{FCCS.C}$ (m/min)	2.4	4.8	7.3	7.7

Figure 1 summarizes the sensitivity of the torching potential, *TP*, to the fuel strata gap, *FSG*. The results indicate that Fuelbed 164 (circles) consistently has higher *TP* values than Fuelbed 282, likely due to the lower *CBH* and $R_{FCCS.S}$. Furthermore, consistent with Table 2, these stands only crown when the *FSG* is relatively low (<1m), and under extreme fire weather conditions.

Lastly, the sensitivity of the crown fire parameters to the reaction intensity was tested. Reports in the literature suggest that Aleppo pine stands are considerably more flammable than typical fuels in North America [22]. Based on this observation, fire behaviour for a North American fuel model #10 (timber litter, heavy dead-and-down fuel loading) was increased four times to better represent the more flammable fuel conditions in Mediterranean Aleppo pine stands [22]. Following this lead, the reaction intensity in FCCS (I_R, eqn. 4) was also increased four times. With this adjustment, both stands have the potential to actively crown in all four fire weather classes (data not shown). Moreover, $R_{FCCS.C}$ increased four-fold to ranges of from 3 to 12 m/min, and from 10 to 30 m/min, for Fuelbeds 164 and 282, respectively (Figure 2). $R_{FCCS.S}$ values ranged from 7 to 12 m/min depending on fire weather class, but with much smaller differences between fuel models (< 20%).

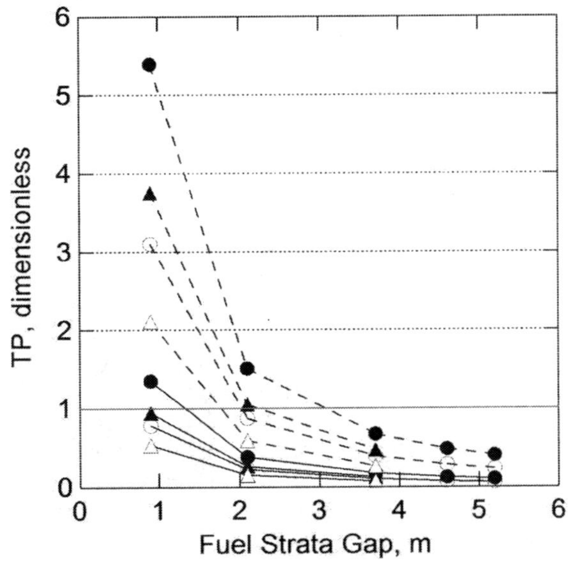

Figure 1: Torching potential (*TP*) as a function of the fuel strata gap, the FCCS fuelbed number, fire weather class, and fire intensity adjustment factor. FCCS Fuelbed 164 (circles) and 282 (triangles) are shown for moderate fire weather (open symbols) and extreme fire weather (closed symbols). *TP* values without the factor-of-4 adjustment for I_R are shown as solid lines and with the factor-of-4 adjustment as dashed lines. *TP* values greater than 1 indicate the potential for crown fire, either active or passive.

4 Conclusions and recommendations

The FCCS crown fire potential equations yielded crown fire spread rates ranging from 2 to 7 m/min for the moderate to extreme fire weather classes, respectively (see Table 2). These crown spread rates are similar to those reported for typical Aleppo pine stands in Greece (1 to 5 m/min [18]), and observations in Maritime pine (*Pinus pinaster*) stands in Portugal (2 to 4 m/min [23]). Furthermore, in the moderate fire weather class, the FCCS crown fire spread rates (2 to 5 m/min) compared favourably with observed crown fires in Maritime pine reported in [15, Table 4] under similar fire weather condition; specifically, 2 to 4 m/min.

Adjusting for the differences in flammability between Aleppo pine and fuels in North America [22] resulted in an increase in the crown fire potentials, *TP* and *AP*, as well as the predicted $R_{FCCS.C}$ values (Figures 1 and 2). The resulting $R_{FCCS.C}$ values are considerably lower than those reported for CFIM/CFIS model runs for these stands [19]. However, in a comparison of models, Scott [24] reported that CFIM provided high estimates of both crown fire initiation thresholds and crown fire spread rates.

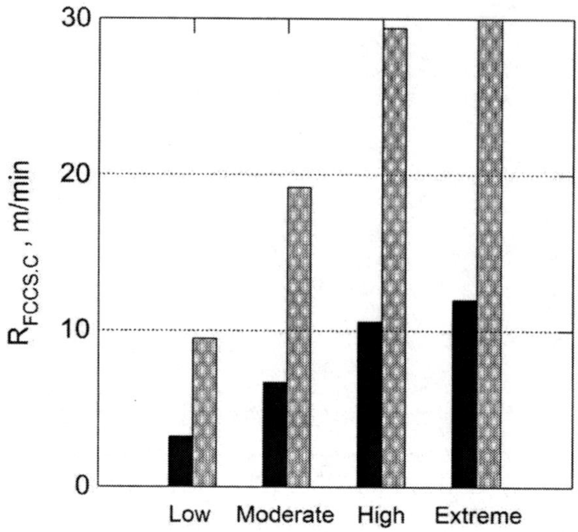

Figure 2: $R_{FCCS.C}$ (m/min) for FCCS Fuelbeds 164 (black) and 282 (hatched) as a function of fire weather class, assuming a four-fold increase in flammability.

The onset of crown fire in FCCS, indicated by *TP* and *AP*, was shown to be dependent on several factors, including: fire weather class, the presence of surface and ladder fuels, *CFL,* and *CBD.* These findings are consistent with recommendations to reduce wildfire effects in Mediterranean Maritime pine stands, such as reducing surface fuel loading, removal of ladder fuels, and reducing crown fuel loading [25]. In FCCS, the crown fire predictions are influenced proportionately by the adjustment factor for reaction intensity. While we used an adjustment factor of 4, adjustment factors of 1.5 to 2 are sufficient to initiate crown torching in the modelled FCCS fuelbeds.

The good correspondence with observations and modelling results in Aleppo pine suggest that the FCCS crown fire potentials might be a useful tool for fire managers in the Mediterranean region to consider when evaluating the relative behaviour of crown fires in vegetated canopies.

References

[1] Schaaf, M.D., M.D. Schreuder, Sandberg, D.V., and Riccardi, C.L., Fire potential rating for wildland fuelbeds using the Fuel Characteristic Classification System, *Can. J. For. Res.* **37**, pp. 2464-2478, 2007.

[2] Van Wagner, C.E., Conditions for the start and spread of crown fire, *Can. J. For. Res.* **7**, pp. 23-34, 1977.

[3] Rothermel, R.C., Predicting behavior and size of crown fires in the Northern Rocky Mountains, USDA For. Serv. Res. Pap. INT-438, 1991.

[4] Rothermel, R.C., A mathematical model for predicting fire spread in wildland fuels, USDA For. Serv. Res. Paper INT-115, 1972.

[5] Sandberg, D.V., Riccardi, C.L., and Schaaf, M.D., Reformulation of Rothermel's wildland fire behavior model for heterogeneous fuelbeds, *Can. J. For. Res.* **37**, pp. 2438-2455, 2007.

[6] Ottmar, R.D., Sandberg, D.V., Riccardi, C.L., and Prichard, S.J., An overview of the Fuel Characteristic Classification System (FCCS)–quantifying, classifying, and creating fuelbeds for resource planning, *Can. J. For. Res.* **37**, pp. 2383-2393, 2007.

[7] Scott, J.H., and Reinhardt, E.D., Assessing crown fire potential by linking models of surface and crown fire behavior. *USDA For. Serv. Res. Pap.* RMRS-RP-29, 2001.

[8] Stocks, B.J., Alexander, M.E., and Lanoville, R.A., Overview of the International Crown Fire Modelling Experiment (ICFME), *Can. J. For. Res.* **34**, pp. 1543-1547, 2004.

[9] Butler, B.W., Finney, M.A., Andrews, P.L., and Albini, F.A., A radiation-driven model for crown fire spread, *Can. J. For. Res.* **34**, pp. 1588-1599, 2004.

[10] Cruz, M. G., Alexander, M. E., and Wakimoto, R. H., Definition of a fire behavior model evaluation protocol: A case study application to crown fire behavior models. In: *Fire, Fuel Treatments, and Ecological Restoration: Conference proceedings*. USDA Forest Service Proceedings RMRS-P-29, pp. 49-67, 2003.

[11] Cruz, M.G., Alexander, M.E., and Wakimoto, R.H., Assessing the probability of crown fire initiation based on fire danger indices, *For. Chron.* **79**, pp. 976-983, 2003.

[12] Cruz, M.G., Alexander, M.E., and Wakimoto, R.H., Modeling the likelihood of crown fire occurrence in conifer forest stands, *For. Sci.* **50**, pp. 640-658, 2004.

[13] Cruz, M.G., Butler, B.W., Alexander, M.E, and D.X. Viegas, Development and evaluation of a semi-physical crown fire initiation model. In: Viegas, D.X. (Ed.), *Proceedings of V International Conference on Forest Fire Research*, Millpress Sci. Publ., Rotterdam, Netherlands, CD-ROM, 2006.

[14] Cruz, M.G., Butler, B.W., Alexander, M.E, Forthofer, J.M., and Wakimoto, R.H., Predicting the ignition of crown fuels above a spreading surface fire, Part I: model idealization, *Int. J. Wildl. Fire.* **15**, pp. 47-60, 2006.

[15] Cruz, M.G., Butler, B.W., and Alexander, M.E., Predicting the ignition of crown fuels above a spreading surface fire. Part II: model evaluation. *Int. J. Wildl. Fire.* **15**, pp. 61-72, 2006.

[16] Alexander, M.E., Cruz, M.G., and Lopes, A.M.G., CFIS: A software tool for simulating crown fire initiation and spread. In: Viegas, D.X. (Ed.), *Proceedings of V International Conference on Forest Fire Research.* Millpress Sci. Publ., Rotterdam, Netherlands. CD-ROM, 2006.

[17] Alexander, M.E., and Cruz, M.G., Evaluating a model for predicting active crown fire rate of spread using wildfire observations, *Can. J. For. Res.* **36**, pp. 3015-3028, 2006.

[18] Dimitrakopoulos, A., Mediterranean fuel models and potential fire behavior in Greece, *Int. J. Wildl. Fire* **11**, pp. 127-130, 2002.

[19] Mitsopoulos, I., and Dimitrakopoulos, A., Canopy fuel characteristics and potential crown fire behavior in Aleppo pine (*Pinus halepensis*) forests, *Ann. For. Sc.* **64**, pp. 287-299, 2007.

[20] FCCS, version 2.0, www.fs.fed.us/pnw/fera/fccs/

[21] Riccardi, C., Ottmar, R., Sandberg, D., Andreu, A., Elman, E., Kooper, K., and Long, J., The fuelbed: a key element of the Fuel Characteristic Classification System, *Can. J. For. Res.* **37**, pp. 2394-2412, 2007.

[22] Carmel, Y., Shlomit, P., Jahashan, F., and Shoshany, M., Assessing fire risk using Monte Carlo simulations of fire spread, *For. Ecol. and Mgmt.* **257**, pp. 370-377, 2009.

[23] Fernandes, P., Loureiro, C., and Botelho, H., Outcomes of a high-intensity experimental fire in a maritime pine stand. In: *Proceedings of the International Scientific Workshop on "Forest Fires in the Wildland-Urban Interface and Rural Areas in Europe: An Integral Planning and Management Challenge."* May 15-16, 2003, Athens, Greece, 2003.

[24] Scott, J., and Burgan, R., Standard fire behavior fuel models: a comprehensive set for use with Rothermel's surface fire spread model, *USDA Forest Service General Technical Report* RMRS-GTR-153, 2005.

[25] Fernandes, P., and Rigolot, E., The fire ecology and management of maritime pine (*Pinus pinaster* Ait.), *For. Ecol. and Mgmt.* **241**, pp. 1-13, 2007.

Author Index

Abt K. L. 197
Amorim J. H. 83
Aranda J. M. 15, 119
Azari H. 233

Batista A. C. 3
Benavente D. 143
Borrego C. 71, 83
Butry D. T. 197

Carrera F. 57
Cascão P. 83
Chas-Amil M. L. 247
Chávarri L. 119
Chetehouna K. 221
Conde F. C. 149
Conti M. 37
Cordeiro C. R. 83
Corgnati L. 163, 173
Couce E. 47
Courty L. 221

de Castro A. J. 15
Diez C. 15

Essen H. 97

Fazenda A. L. 149
Fernández-Gómez I. 15
Ferreira A. J. 83
Freitas S. 149

Garo J. P. 221
Goossens R. 107
Guerin S. 57
Guijarro M. 15

Heinen S. 97
Hernando C. 15

Jankovic L. 25

Kirsch A. 187
Knorr W. 47
Koutitas G. 25
Kruell W. 97

Lalkovič M. 131
Lhermitte S. 107
Longo K. 149
López F. 15, 119
Losso A. 163, 173
Lucio P. S. 149

Madrigal J. 15
Marconi U. M. B. 37
Martins V. 71, 83
Meléndez J. 119
Miranda A. I. 71, 83
Monteiro A. 71
Moreira D. S. 149

Pajtíková J. 131
Pavlidou N. 25
Pereira J. F. 3
Perez J. L. 71
Perona G. 163, 173
Pita L. P. 83
Prestemon J. P. 197, 247

Ramos A. M. 149
Ribeiro L. M. 83
Rideout D. B. 187, 207

Sá E. 71
San José R. 71
Sandberg Da. V. 259
Schaaf M. D. 259
Schaap M. 71
Schreuder M. D. 259
Silva A. M. 149
Soares R. V. 3

Tavares R. 83
Tchepel O. 83

Tobera R. 97
Touza J.................................... 247

Valente J. 83
Veraverbeke S.......................... 107
Verstraeten W. 107
Viegas D. X. 83, 221

von Wahl N............................... 97

Wei Y.............................. 187, 207
Willms I. 97

Ziesler P. S............................... 207